中国金鱼图鉴

GOLDFISH OF CHINA

DESCRIPTIONS AND ILLUSTRATIONS
OF DIVERSED GOLDFISH
IN CHINA

主编:叶其昌 曲利明

海峡出版发行集团 | 海峡书局
THE STRAITS PUBLISHING & DIBLISHING GROUP

图书在版编目（CIP）数据

中国金鱼图鉴 / 叶其昌，曲利明主编 . -- 福州 ：
海峡书局，2017.1
ISBN 978-7-5567-0261-9

Ⅰ．①中… Ⅱ．①叶… ②曲… Ⅲ．①金鱼－中国－
图集 Ⅳ．① S965.811-64

中国版本图书馆 CIP 数据核字（2016）第 238528 号

出 版 人：林 彬
主　　编：叶其昌　　曲利明
撰　　文：曹 峰
责任编辑：陈 婧　　林前汐　　廖飞琴　　卢佳颖
装帧设计：董玲芝　　李 晔　　黄舒堉
封面设计：黄硕燊
插　　图：黄宏宇　李 晔
翻　　译：Jasmine Zhu　　Phoebe Xu　　Azure Zheng

ZHŌNGGUÓ JĪNYÚ TÚJÌAN

中国金鱼图鉴

出版发行：海峡出版发行集团 海峡书局
地　　址：福州市鼓楼区五一路北段 110 号 1# 楼 9 层
邮　　编：350001
印　　刷：深圳市泰和精品印刷有限公司
开　　本：889mm × 1194mm　　　1/16
印　　张：33.75
图　　文：540 码
版　　次：2017 年 1 月第 1 版
印　　次：2017 年 1 月第 1 次印刷
书　　号：978-7-5567-0261-9
定　　价：480.00 元

《中国金鱼图鉴》编委会

◎ **编委会成员** /（排名不分先后，按姓氏笔画排列）

杨小强　何为　张正农　陈钰　陈镇平　林海　黄宏宇　曹峰

◎ **摄影** /（排名不分先后，按姓氏笔画排列）

叶其昌　曲利明　何为　沈宁　张正农　林海　黄宏宇　曹峰　鲍华

特别鸣谢

◎ **协作单位** /（排名不分先后，按姓氏笔画排列）

九亭金鱼	武汉中秋金鱼养殖场
广东清远市清新区太和镇五星金鱼养殖场	南京金鱼俱乐部金鱼养殖场
中国兰寿网北京鱼友会	闽侯县荆溪关中潘氏观赏鱼养殖场
长乐市首占福连观赏鱼养殖场	闽侯县南屿爱民金鱼场
长乐鹤上龙峰金鱼场	闽侯县南屿旗山潘墩金鱼场
长乐鹤上春之初金鱼养殖场	闽侯县南通春园鲤生态养殖场
东海水族有限公司	闽侯县南通起荣金鱼养殖场
北京汉鳞汇金鱼文化苑	闽侯县洲头观赏鱼科研养殖场
北京思漫鱼场	闽侯县旗山林雕金鱼养殖场
北京顺民义友观赏鱼养殖基地	闽侯洋润水族科技发展有限公司
如皋市长蝶观赏鱼专业合作社	唐山中天鱼场
如皋金鑫观赏鱼有限公司	浙江高峰金鱼展示场
如皋蝶舞观赏鱼养殖场	漳州市闽南渔村
苏州保发金鱼场	

◎ **协作鱼友** /（排名不分先后，按姓氏笔画排列）

刘振勇（候鸟）	秦峰（渔公渔婆）
刘淼（五花皮球）	袁宏伟（武汉老袁）
齐文远	高溯源
汪聿钢	郭栋（aggoom）
沈宁（五木）	黄之旸

序

　　金鱼发源于中国，是自然界的野生鲫的突变体经中国养鱼人和赏鱼人的发现和挑选以及长期的豢养、选育而分化出来的人工品种，据史料记载已有近1800年的历史。金鱼的演化过程中，因为人类在遗传育种上的干预，使不存在生殖隔离的同一物种，分化出千姿百态、形态各异的几百个金鱼品系，这种现象在鱼类演化和分类学上十分罕见，具有很高的研究价值。

　　在古代中国，最初被人们放生的红色鲫鱼，承载了中国人对生命的敬畏和对美好生活的向往，体现了中国文化中珍爱生命、善待天地万物的思想。直到南宋（1127-1279），金鱼由池养改为盆养，进一步缩短了与人的距离，人们可以更仔细地观察不同个体之间细微的差异，然后通过人工繁殖，挑选自己认为有美好表现的个体留下饲养，再繁殖下一代。

　　根植于中国文化土壤的金鱼，也同样得到了全世界人民的喜爱。20世纪50年代，周恩来总理多次把金鱼作为国礼赠送给国际友人。

　　1502年中国金鱼传入日本，17世纪末传入英国，18世纪中叶又传到英国以外的欧洲其他地区，1874年传入美国……到今天，金鱼已经传遍全世界，在世界各地生根发芽。世界各地人民根据自己对美的理解，繁殖、饲养、选育，逐渐形成了各具品相特色的金鱼品种。

　　和人类家化驯养的其他动物不同，金鱼是人类为了满足自己精神文化层面的需要，以自然物种鲫鱼为原材料，通过遗传学手段塑造出来的人工鱼类品种，其形态特征代表了不同地区、不同时代审美的差异，具有丰富的社会人文内涵。可以说，金鱼是完全依赖于人类精神文化需要而培育的观赏鱼品种。金鱼外部形态的变异，也使得金鱼难以再适应自然水域，只能生活在人类社会环境中，为人类所豢养。由于金鱼遗传变异的特殊性，可以预见，未来必将出现更多形态各异的金鱼品种。

　　《中国金鱼图鉴》集中了一批金鱼研究和养殖的专家，对大量历史资料进行收集整理，走访调查了北京、天津、湖北、江苏、上海、福建、广东等目前金鱼养殖较集中的产区中上百家鱼场，拍摄了数以万计的金鱼照片，基本摸清了我国目前饲养的金鱼品种，从中挑选具有代表性的金鱼257个品种，近800张照片，编入图书，旨在绘制出目前中国金鱼养殖品种的版图，记录下21世纪初中国各地养殖金鱼的品种形态特征，为以后进一步的金鱼研究提供了重用依据。

　　《中国金鱼图鉴》还重新梳理了目前中国金鱼的品种分类脉络，明确了以形态特征为主的命名规则，相信对改善目前市场上对于金鱼分类众说纷纭的局面有所帮助，是金鱼养殖生产、贸易、鉴赏的一本重要工具书。

　　有感于中国金鱼的精妙，《中国金鱼图鉴》的精美，《中国金鱼图鉴》作者们的精神，作为一名长期痴迷于中国金鱼千变万化、品种众多的鱼类遗传育种学家，钦慕之情溢于言表，故为序。

Preface

Goldfish, originated from China with a history of 1800 years, has been domesticated and cultivated by Chinese goldfish raisers and fanciers from wild carp after their variations. The evolution, interacted by man, creates hundreds of varieties of goldfish in different shapes and colors. This is rare on goldfish evolutionand classfification and of great value on fish classification studies.

In ancient China, the red wild carp were freed due to man's worship of lives and great wish for life, reflecting Chinese culture of life-centered and mercy to nature creatures. By the South Song Dynasty (1127-1279) , goldfish is changed from pond raise to bowel raise, make it getting closer to man. The raisers thus can observe the detailed differences between each variety, and with selective breeding, develop those of good characteristics and reproduce them.

Goldfish, rooted in culture of China, is well received all over the whole world. In 1950s, Premier Zhou has sent to international friends with goldfish as presents for many times.

Chinese goldfish has been introduced to Japan in 1502 and to England at the end of 17th century, to other regions of Europe in mid. 18th century, and to America in 1874… By today, they can be found almost everywhere in the world. People all over the world reproduce, cultivate, selective breed goldfish per their own aesthetics values, thus many different colors and shapes of goldfish have been formed.

Unlike other domesticated animals, goldfish is the spiritual pursuit for man. Deriving from the wild carp, it has been artificially developed into all kinds of varieties, which has reflected the difference on regions and times, with abundant social culture significance included. In fact, goldfish is an ornamental fish variety that lives for the man's mental satisfaction. Due to the variations, goldfish have merely not able to live in the natural water areas, instead, they depend more on artificial breeding, then got survived in human's society. It is foreseeable that in the future, with the unique genetic variations, more and more varieties will be come into people's lives.

"Goldifsh of China" has collected a large deal of historical records, and researches from many goldfish learners and goldfish raisers; researchers have visited and investigated many varieties from places as Beijing, Tianjin, Hubei, Jiangsu, Shanghai, Fujian, and Guangdong. Ten thousand of photos have been taken. They have clearly understood the main varieties, 257 of which with up to 800 pictures have been edited in this book, with the aim to draw the contemporary map of Chinese goldfish so as to provide an important foundation for futher researches.

The book "Goldfish of China" has also reorganized the goldfish classification, establishing a naming principle based on morphological characters. It is believed with the help of this book, the standard of goldfish classification can be set up. It will be served as an important guide for goldfish cultivation, trade and for appreciation.

Impressed by the delicacy of Chinese goldfish as well as the fineness of this book, as a Genetic Scientist of goldfish, I wrote the preface out of great admiration to the authors and editors of this book.

oldfish

目 录
CONTENTS

蛋形 > 头型变异类 > 龙头型---400
Egg-Fish > Head Variation > Dragonhead

中国金鱼的起源与演变

文 / 何 为

一、总论

"灰瓦白墙青砖地，水井鱼缸芍药花……"走进古朴的中国传统民宅，常常能够见到这样的景象：点点线线的阳光从树枝间散落下来，撒在鱼缸的一泓清水上格外耀眼。缸里几尾金鱼若隐若现，或低头觅食，或抬头摆尾，与水面的光斑和婆娑的树影嬉戏。缸边一位老人，手中捧一把紫砂壶，轻啜一口香茶，回过头，缸里的金鱼磁石般紧紧吸引着老人的目光。老人身边一幼童，双手扒住齐肩高的缸沿，伸出头看着缸里的金鱼，不时向老人问这问那……这样的生活场景中，到处渗透着圆柔、温和、随意、从势的中国文化气韵，体现出人与环境和谐相处、浑然天成的生活哲学。其中，金鱼与人们生活的密切程度可见一斑。而金鱼本身就是中国传统文化的产物，其中蕴含了中国人独特的审美情趣和生活态度。

类似从野生动物驯化而来的家禽家畜，金鱼是由野生鲫鱼进化而来的，直到最终形成今天繁杂的族系，金鱼已经完全脱离了自然生态食物链，不再能够适应野生自然环境而必须依赖于人类生存。不同的是，人类驯养家禽家畜的最初动机大多带有护卫、肉用等实用价值，而金鱼，是由最初人类对自然物种的崇拜逐渐过渡到后来为了满足自身精神需要而豢养。因此，金鱼的外观、颜色、游姿等等，在一定程度上反映了当地当时的文化背景和审美要求。例如，中国金鱼中的龙睛，由于眼睛凸出，形似中国传统图腾龙的形象而大受欢迎，但没有中国文化背景的地区觉得龙睛的眼睛凸出，形象怪异，不容易接受；中国的蛋形金鱼传播到日本后，受到日本文化和审美的影响，逐渐演变出的日寿粗壮雄浑，吻凸极为发达，为了达到身体的平衡而尾部卜弯，和中国类似的蛋形狮头、虎头具有明显的区别。而兰寿、狮头、虎头都是由同一祖先——原始的蛋形金鱼演化而来。日寿引入中国以后，又分成两支，一支严格遵从日本的筛选标准，保留了日寿的各项特征，成为在中国饲养的"日寿"；另外一支，受到中国文化背景的影响，形成新的体态特征，更加圆润，和原本的日寿具有明显区别，在短短几十年时间里演化出中国兰寿。

不同的文化背景演化出具有不同特征的金鱼。据记载，中国金鱼1502年传入日本。日本有着深厚的鱼文化背景，由于其地理位置和生活方式，鱼已经深入到日本人生活的方方面面。金鱼传到日本以后，逐渐演化出具有日本特色的日本金鱼体系，无论文形还是蛋形，都与当时刚刚传到日本的中国金鱼有着很大的区别。17世纪末金鱼传到英国，经过300多年的演变，至今英国最著名的金鱼品种布里斯托金身上只保留了单尾心形尾，已不再像中国和日本金鱼般重视头型、眼型等变异。

金鱼和人类文化背景有着这样密切的关系，因此，金鱼不仅仅是单纯的生物学"品种"，而是和人类精神文明密不可分的"艺术物种"。作为自然物种的"金鲫"，摆脱了自然生态食物链，进入人类生活，依赖人类而生存、繁衍和演化后，才成为真正意义上的"金鱼"。研究金鱼，既离不开金鱼的生物学属性，也离不开人类社会文化背景和人类的审美情趣。因此本书在图片拍摄选择的时候，不仅仅遵照生物学分类的标准，同时也力求图片能够体现各个品种金鱼不同的美。

值得关注的是，经过数百年的变异、杂交和人工选择，金鱼形态特征的变化呈现出加速的趋势。新的突变和人为杂交使新的性状出现的频率更高。同时，人类的审美标准也随着时代而变化，反映在金鱼的演化上，新的形态特征不断涌现，一旦这些形态特征和原有的特征具有明显区别，并且可以稳定地遗传到下一代身上，就会被认为是一个新的品种。

随着生物科技的飞速发展，新的育种手段应用到金鱼培育上，新品种的形成周期越来越短，产生速度越来越快，金鱼的品种变化将持续下去，金鱼这一人类文明的奇葩将会变得愈加多姿多彩。而本书，希望通过对21世纪初中国金鱼的历史演变、分类观点、产地的记录，特别是通过图片记录金鱼的形态特征，描绘出当代中国金鱼的版图，成为金鱼演化过程的一块里程碑。

二、本书的主要观点

金鱼起源于中国，由野生鲫鱼演化而来，世界各地的金鱼都是由中国传播出去的。

金鱼脱离自然野生环境，演化出各种形态性状，已经不能适应野生自然环境。这种演化现象的基础，是人类的豢养行为，其形态特征的形成和人文背景分不开。因此，真正意义上的金鱼出现是从家养时代开始，即我国的南宋时期（1127—1279）。

金鱼演化和品种形成的原因是由于突变、杂交和人工选择3个因素造成的。在这3个因素中，突变是基础，杂交和人工选择是手段。

金鱼与野生鲫鱼不存在生殖隔离，染色体组型与中国野生鲫鱼相同。因此，把金鱼定义为鲫鱼的变种，金鱼的学名为 *Carassius auratus* var.

金鱼的动物分类学分类地位：

脊索动物门

  脊椎动物亚门

   有颌总纲

    辐鳍鱼纲

     新鳍鱼亚纲

      真骨鱼组

       正真骨鱼亚组

        骨鳔鱼总目

         鲤形目

          鲤科

           鲤亚科

            鲫属

金鱼各族群、品系之间不存在生殖隔离，系统发育和胚胎发育都没有明显的不同。而且染色体、同工酶的研究结果表明各族群、品系之间缺少显著差异。另一方面，金鱼的形态特征差异又极大，各种形态差异存在中间过渡类型。和其他鱼类相比，金鱼各项特征在后代的表现存在分化现象，所谓"稳定的遗传"不同于动物学分类上的稳定遗传，是指亲代特征在子代得到表现的比例相对较高。因而，严格的动物学分类方法和标准在金鱼上并不适用，需要另外选择适用于金鱼的分类体系，在同一个生物学"种"的定义下，对金鱼进行分类和命名。金鱼"品种"的含义和生物学"种"的含义是不同的。

本书以金鱼最原始的表现类型"文鱼"为原点，把各种金鱼发生变异的性状特征与"文鱼"相比较，作为分类依据，采用"二分法"对金鱼进行分类描述。

金鱼的很多特征呈现数量遗传的现象，而不是"非此即彼"，因此很难用鱼类形态学分类对金鱼品种进行界定。本书以某个品种形态特征是否能够"稳定遗传"作为界定金鱼品种的主要标准。

值得注意的是，金鱼的品种特征，很少在后代中呈现百分百的遗传，所谓"稳定遗传"只能是较高比例的遗传。而具体的量化指标，目前尚未有人进行系统研究。因而目前金鱼并没有切实可用的考种标准。本书所列出的金鱼品种，主要根据已经被大多数金鱼饲养者约定俗成认可的品种，结合遗传比例的调查结果已确定。

目前普遍接受的金鱼命名大多是根据金鱼性状特征来命名，除了应用在商业上的商品名以外，还有很多具有人文内涵的名称。本书在明确提出新的分类体系基础上，还是以约定俗成的传统命名作为正名（如鹤顶红、寿星等等），文学命名（如双灯照雪）或别名（如点红）只在细述中略作交代。

需要特别说明的是，金鱼的品种变化是绝对的，稳定只是相对的，随着种类的积累，用于杂交的种类越来越多，新的特征会在金鱼身上不断出现。新的特征是否会被保留下来，取决于人工选择，而人工选择的标准会随着不同地区、不同时代的审美而改变。目前金鱼新品种的形成呈加速的趋势。

本书在提出分类体系和命名规则时，兼顾到分类体系和命名规则可以应用到新的特征和品种，为金鱼的分类和命名找到统一的可延续的方法。

在中国传统金鱼中，翻鳃作为一个变异特征而界定了一个大的族群，在头型变异类、尾鳍变异类等各个特征变异的族群里都曾经有翻鳃金鱼的记录。但随着时代的发展，对金鱼的审美有了很大变化，翻鳃金鱼逐渐被视作残疾特征而舍弃，今天已经很少有人专门培育翻鳃金鱼了。因此，本书没有把翻鳃列入一个变异特征加以区分品种。这也是为什么2007年金鱼有记载330多个品种，而本书只收录了200多个品种的主要原因。

《中国金鱼图鉴》收录的金鱼图片均在中国拍摄，书中出现的"日寿""东锦""地金""南京"等品种的金鱼拍摄对象均为已经引入中国，并在中国繁殖饲养的金鱼，因此也纳入"中国金鱼"的范畴，沿用原产地名称是为了表达对金鱼品种创新者的敬意。

三、金鱼的起源和演变

金鱼起源于中国，由中国的野生鲫鱼演化而来，是中国人根据自己对美的理解，以自然物种为基础材料，采用最基本的遗传育种手段塑造出的人工养殖品种。其外部形态色彩变化之多，与野生原种差异之大，在世界上的动植物中几乎是绝无仅有的。

自然环境中一个野生物种的形成，普遍认为需要数万年的时间，人类驯养野生动物，也需要数千年。即使是家禽中人类驯养时间最短的家鸡，也有6000年历史。而野生鲫鱼演化出千姿百态的金鱼，仅用了区区不到千年时间，这不得不说是一个奇迹，体现了人类对自然物种遗传干涉的作用。

纵观从鲫鱼到金鱼的演化过程，可以分为5个阶段。

在中国的野外自然环境中，存在红黄色野生鲫鱼。1981年5月，王春元曾在宁夏发现红色、白色、蓝色和红白相间的彩色鲫鱼。这些彩色鲫鱼虽然在自然界难得一见，但偶尔还是可以遇到的。中国古代，把这类特殊颜色的鲫鱼称为"文鱼"，以区别于普通的"粗野"鲫鱼。

古代中国文化中素有对自然界异形异色动物因敬畏而放生的现象。这种现象自汉代佛教传入中国，成为本土化的宗教以后，与佛教中"慈悲为怀，不杀生"的信条不谋而合，放生之风日盛。

在这个阶段，记载中的"文鱼"更多的是作为一种人们对自然的观察记录而存在，人们只是把野生的红黄色鲫鱼放养到放生池中，偶尔投以食料，寄托自己对自然物种的敬畏和仁爱，所述的"文鱼"和人类生活的关系与后来的"金鱼"有着很大的差别，此时的"文鱼"，虽然与金鱼分类上所说的"文鱼"字面相同，其含义是完全不同的。所谓的"金鱼""赤鳞鱼"也只是颜色呈黄色、红色，其形状与野生鲫鱼完全相同。

因为这个阶段杭州、嘉兴一带放生池中红黄色鲫鱼放养较多，对红黄色鲫鱼的记载也较详细，因而浙江杭州、嘉兴被视为金鱼的发源地。

公元10世纪，宋朝移都临安，史称南宋。南宋皇帝赵构退位后，在德寿宫中设池，命人四处收集金银鱼带回饲养。金鱼自赵构始，走入皇宫。由于皇帝的喜爱，王公大臣对金鱼趋之若鹜，豪庭贵宅纷纷辟出池塘饲养金鲫鱼。

红黄色鲫鱼由放生池中的半家养形式过渡到家池饲养，金银色的鲫鱼开始真正走进了人们的生活。人们开始不仅仅为了不杀生而放鱼入水自生，出现了为了赏玩而饲养、照顾，为了盈利而人工饲养繁殖；由从野外收集放养到放生池中，变为直接在家池中豢养，前者是放入池中让其自己生存，而后者则是人工饲养，投喂饲料。

南宋以后越来越多的人开始用盆缸养育金鱼，直到明朝后期，金鱼的饲养方式逐渐从池养为主过渡到盆养为主。饲养方式的改变，对金鱼的演化和品种的形成起到了巨大作用，直接导致了金鱼演化史上第一个新品种形成高峰期的出现。

中国金鱼家化史和遗传学研究的奠基人陈桢先生充分考察中国史料以后注意到，1163年到1267年的104年期间，金鱼的饲养方式以家池饲养为主，此期间，金鱼只出现了白色、花斑色两个品种，而体形变异未见记载。盆养时代开启以后，1547年到1643年的96年间，盆养已经成为了金鱼饲养的主要方式，这个阶段新记载的品种多达10个。

这段时期，人们虽然能够更加仔细地观察、挑选金鱼，但这种挑选集中在对金鱼的形态表现的挑选，还没有把金鱼的形态和幼鱼的选择以及亲本的选择结合起来。

明末清初以后，关于金鱼饲养和挑选的记载逐渐多了起来。盆养金鱼，使人们得以近距离更加仔细地观察金鱼，从而以更多的细节区分品种。这说明人们已经开始注意到金鱼从小到大的变化，意识到金鱼从胚胎发育，孵化山膜的时候形态、颜色完全相同，到分化出形态色彩各异的不同品种，是在生长、发育过程中完成的。不同品种的培育，需要从小开始进行人工选择，而这种选择，使得偶然发生的变异得以在幼鱼阶段就被保留下来，直至饲养到成鱼以后完全展示出来。

金鱼饲养的关键技术，一是人工繁殖并养大，二是品种特征的培育。前者是基础，后者是核心。金鱼之所以成为金鱼，就是在于其不同于其他鱼类的品种特征具有文化内涵和审美价值。当人们开始认识到从幼鱼时期完全相同的形态色彩会逐渐变化成为以后不同的品种特征时，挑选幼鱼成为了饲养金鱼最为重要的环节之一。而挑选的标准，在不同地区、不同时代，是不一样的，带有强烈的主观因素。因为主观因素对人工选择的影响，从而导致了不同时代的文化背景和审美情趣在金鱼身上得以体现，也就是金鱼的文化内涵来源所在。直到今天，金鱼的挑选标准和方法，在不同鱼场有着巨大的差异，成为鱼场培育金鱼的主要"秘技"之一。

在人工选择阶段，人们是从幼鱼阶段选择鱼苗。伴随着对鱼苗的人工选择，人们开始注意到亲本对后代的影响，从而开始根据自己希望得到的性状，有意识地选择亲本进行杂交。杂交育种是人们改造金鱼的重要工具，使得人们能够进一步根据自己对美的理解塑造金鱼的形态特征。人们对杂交育种技术的掌握，填补了金鱼培育技术体系最后的一块空白，从此以后，"鱼苗挑选——养成——亲本挑选——杂交——人工繁殖"这一金鱼培育的技术路线最终形成，一直沿用至今。一个新的特征一旦形成，通过反复多次的稳定筛选，最终成为一个新的金鱼品种。

The Origin and Evolution of Chinese Goldfish

Article / He Wei

1. Over View

"Gray tile, white wall, blue brick ground, well, aquarium and peony..." Walking into an old traditional Chinese house, you will often see such a scene. Sunlight is cast down branches in dots and rays, and is exceptionally bright on the clear water of aquarium. Several goldfish are looming. Some lower their head to hunt for food,some raise their head and swim, playing with the light spot or tree shadow on water. Next to the aquarium is an old man, holding a Yixing clay teapot in his hand and sipping tea. He turns around, glaring at the goldfish in the water. A little boy standing by the old man clings to the rim of aquarium, which is as tall as his shoulders, looks at goldfish in the aquarium and asks the old man questions from time to time… Such a scene is full of harmony, gentleness, casualness and obedience in Chinese cultural background and reflects the life philosophy of harmony and naturalism between man and environment. The close relationship between goldfish and people's life is evident from the scene. While goldfish itself is a product of traditional Chinese culture and implies the unique aesthetics and attitude towards life of Chinese people.

Similar to poultry and livestock domesticated from wild animals, the goldfish also evolved from wild carp, and finally formed a complex family today. The goldfish is completely divorced from the natural ecological food chain. They no longer adapt to the wild natural environment, but survive on human beings.The difference is that originally, human beings domesticated poultry and livestock to guard themselves or eat their food, i.e., for practical value. At the beginning, the goldfish was raised out of people's worship for natural species. But later, it was raised to satisfy people's spiritual needs. Therefore, the appearance, color, action and gesture, etc. of goldfish, to a certain extent, reflect the local and cultural background and aesthetic needs. Dragon Eyes in China, for example, is well-received because its protruding eyes resemble the image of traditional Chinese dragon. But in areas without a Chinese cultural background, it seems weird and unacceptable. After Chinese Egg-Fish was introduced to Japan, influenced by the Japanese culture and aesthetics, it gradually evolved into Japanese Ranchu, a sturdy and strong species with a developed snout. In order to achieve the balance of body, its tail was bent. It was distinct from Chinese Lionhead in Egg-Fish and Tigerhead, while the three of them evolved from the same ancestor-ancient Egg-Fish. After the Japanese Ranchu was reintroduced to China, it was divided into two groups. One group strictly complied with the screening standard of Japan, retained the features of Japanese Ranchu and became China-bred "Japanese Ranchu". The other group was influenced by Chinese cultural background and formed a new body shape. They were more rounded and distinct from the original Ranchu. Chinese Ranchu evolved within just a few decades.

Different cultural backgrounds bred goldfish with different characteristics. According to documentation, Chinese goldfish was introduced into Japan in 1502[8]. Having a prolonged history of fish and with special geographical location and lifestyle, the fish has reached deep into every aspect of Japanese people's life. After the goldfish was introduced to Japan, it gradually evolved into a Japanese goldfish system with Japanese characteristics, either Fantail Goldfish or Egg-Fish were quite different from the earliest Chinese goldfish introduced to Japan. In the late 17th Century, the goldfish was introduced to Britain. After evolving for 300 years, the most famous goldfish variety in Britain today, Bristol goldfish, not like Chinese and Japanese goldfish which are favored on head and eye variation, only retains a single heart-shaped tail.

Due to the close relationship between goldfish and human culture background that goldfish is not only a simple "variety" in biology, but also an art species that are inseparable from people's spiritual civilization. "Gold carp", as a natural species, breaks away from the natural ecological food chain, enters people's life, survives, reproduces and evolves depending on humans, and becomes "goldfish" in the true sense. The study on goldfish is inseparable from the biological features of goldfish, and also inseparable from the social and cultural background and aesthetics of human beings. This is why this book not only follows a biological classification standard, but also strives to show the beauty of different goldfish varieties by taking and choosing pictures.

It is noteworthy that after hundreds of years of variation, hybridization and artificial selection, the morphological characteristics of goldfish change even more quickly. New variations and artificial hybridization make new characteristics appear more frequently. At the same time, aesthetical standards of people also change over time. During the evolution of goldfish, new morphological characteristics emerge constantly. Once these morphological characteristics are distinctively different from the original characteristics, they will be passed on to the next generation steadily and regarded as a new variety.

With the rapid development of biotechnology, new breeding methods have been applied to goldfish cultivation. The formation cycle of new varieties becomes shorter and shorter. New varieties are produced faster and faster. The change of goldfish varieties will continue. The goldfish, as a miracle of human civilization, will become more and more colorful. For this end, this book attempts to depict a contemporary Chinese goldfish map and serve as a milestone in the evolution process of goldfish, by recording the historical evolution, classification and origin of Chinese goldfish in the early 21st Century, especially recording the morphological characteristics of goldfish using pictures.

2. Main Ideas of This Book

The goldfish originated from wild carp in China. All of the goldfish around the world was introduced from China.

The goldfish was divorced from the natural wild environment and developed a variety of morphological characters. It has already been unable to adapt to the wild nature environment. The foundation of such an evolution was human feeding. The formation of the morphological characteristics of goldfish is inseparable from the humanistic background. Thus, the goldfish in the true sense emerged in the domestic age, namely, the Southern Song Dynasty (1127-1279) of China.

The evolution and formation of goldfish varieties are attributed to three factors: variation, hybridization and artificial selection. Among these three factors, variation is the foundation, hybridization and artificial selection are means.

The goldfish and wild carp are not isolated in reproduction. The karyotype of goldfish is the same as that of Chinese wild carp. So the goldfish is defined as a variant of carp. The scientific name of goldfish is *Carassius auratus* var.

The taxonomicstatus of goldfish in zootaxy:

Phylum CHORDATA

Subphylum Vertebrata (Craniata)

Superclass Gnathostomata

Class Actinopterygii

Subclass Neopterygii

Division Teleostei

Subdivision Euteleostei

Ostariophysi

Cypriniformes

Family Cyprinidae (Minnous or Carps)

Subfamily Cyprininae

Carassi

There is no reproductive isolation between different families and strains of goldfish. Their systematic development and embryonic development have no obvious difference. Research findings about chromosome and isozyme indicate that there is no significant difference between all families and strains. On the other hand, the morphological characteristics of goldfish vary greatly. There are also transitional types between different morphologies. Compared with other fishes, various characteristics of goldfish have differential offspring expressions. The so-called "genetic stability" is different from genetic stability in zootaxy. It means that there is a high chance that the parental characteristics are manifested in filial generation. Therefore, strict methods and standards in zootaxy don't apply to goldfish. We need to choose a proper goldfish classification system, classify and name goldfish under the same biological definition of "species". The meaning of goldfish "variety" is different from that "species" in the biological sense.

This book starts with the most primitive goldfish phenotype "Fantail Goldfish", compares the mutated characteristics between goldfish and "Fantail Goldfish", as the basis of classification, and then classifies and describes goldfish using "dichotomy".

Many of the characteristics of goldfish show quantitative heredity, instead of "either-or", so it is difficult to define goldfish varieties using the morphological classification of fish. This book takes whether the morphological characteristics of a certain variety can be "inherited stably" as the main standard of defining goldfish varieties.

It is worth noting that the characteristics of goldfish varieties are rarely 100%

inherited to the offspring. The so-called "stable inheritance" is merely inheritance with a higher proportion. However, the specific quantitative indicators have not yet been studied systematically. Goldfish varieties listed in this book are mainly based on varieties established and approved by most goldfish raisers and determined in combination with the findings about inheritance proportions.

At present, generally-accepted goldfish are mostly named after their characters. In addition to the trade name, there are also a lot of names with humanistic connotations. This book explicitly puts forward a new classification system, applied the traditional naming method like Hedinghong, Lionhead (Southern), the liternal naming like Two lights in Snow or nick name Rudianhong is just breafly described.

It is especially noted that the change of goldfish varieties is absolute, while stability is relative. With the accumulation of varieties, there will be more and more varieties for hybridization. New features will exhibited on goldfish. Whether new features will be preserved or not depends on artificial selection.Nowadays, the criteria of artificial selection will change with aesthetics in different areas and times. The formation of new goldfish varieties tends to accelerate.

When putting forward a classification system and naming rule, this book also considers that the classification system and naming rule must be applied to new features and varieties and intends to find a uniform and sustainable method for goldfish classification and naming.

Among traditional Chinese goldfish, Reversed Gill, as a mutated feature, defines a large group. In head variation, caudal fin variation and other character-variation groups, there were records about Reversed Gill. But with the development of times, goldfish aesthetics have changed a lot. Reversed Gill is gradually dismissed as a symbol of disability. Today, Reversed Gill is rarely specially cultivated. Therefore, this book doesn't list reversed-gill as a mutant feature for distinguishing varieties. This is also why more than 330 varieties were in 2007, but only more than 200 varieties are included in this book.

All the goldfish pictures included in *GOLDFISH OF CHINA* are taken in China. The shooting objects like "Japanese Ranchu" "Azuma Nishiki" "Jikin" "Nankin", and so on, are goldfish introduced to and bred in China. Hence, they are included in the category of "Chinese goldfish", too. To show respect to variety innovators, the goldfish are named after their origins.

3. The Origin and Evolution of Goldfish

Goldfish started from wild carps in ancient China. It was an artificial breed created according to the understanding of Chinese people of beauty, with natural species as the material, using the most fundamental genetic breeding means. It has diverse shapes and colors and distinctively different wild stocks. This is virtually unique among plants and animals in the world.

It is generally believed that the formation of a wild species in natural environment takes tens of thousands of years. Also it takes thousands of years for human beings to domesticate wild animals. Even chickens, which have the shortest domestication time among poultry, have a history of 6,000 years. However, it takes less than one thousand years for wild carp to evolve into goldfish with diverse forms. This is absolutely a miracle and reflects the genetic intervention of human beings in natural species.

The evolution process from carp to goldfish can be divided into five stages.

In the wild natural environment in China, there are red and yellow wild carps. In May 1981, Wang Chunyuan found red, white, blue, red-and-white crucian carps in Ningxia. These color carps were rare in nature, but still seen occasionally. In ancient China, this kind of special color carp was called "Fantail Goldfish" to distinguish from the ordinary "rough" carp.

In ancient Chinese culture, it was very common that out of respect, heteromorphic and heterochromatic animals in nature were freed by people. Since Buddhism was introduced to China and became a local religion, this phenomenon coincided with the Buddhist creed of "leniency and ahimsa" and animal freeing began to prevail.

At this stage, "Fantail Goldfish" was a natural observation record of human beings. People freed wild red and yellow carps into free life ponds and gave them food occasionally, to show their respect and love for natural species. The relationship between such "Fantail Goldfish" and human life was quite different from the subsequent "goldfish". Although at this time, "Fantail Goldfish" was literally the same as in the goldfish classification, the meanings were completely different. The so-called "goldfish" and "scarlet fin soldier" were just yellow and red in color. Their shapes were exactly the same as wild carps.

As many red and yellow carp were freed in free life ponds in Hangzhou and Jiaxing of Zhejiang province, there were detailed records about red and yellow carp, so Hangzhou and Jiaxing were considered the cradle of goldfish.

In the 10th Century, the capital of Song Dynasty was moved to Lin'an and the dynasty was known as the Southern Song Dynasty in history. After the Southern Song emperor Zhao Gou abdicated, he built a pond in Deshou Palace and ordered his men to gather golden carp.

Red and yellow carps changed from semi-domestic in free life ponds to home ponds. Gold and silver carps officially began to enter into people's life. People no longer put fish in water and let them self-fed. Instead, they fed and looked after them to admire their beauty and reproduced them artificially to make profits. Carps changed from wild gathering, free life pond feeding to home pond feeding directly. The former were freed in ponds and self-fed, while the latter was raised by people casting feedstuff.

After the Southern Song Dynasty, more and more people began to use a bowl to raise goldfish. Till the late Ming Dynasty, the breeding way of goldfish gradually changed from pond to bowl. The change of breeding way played a huge role in the evolution and formation of varieties and directly led to the first peak of new varieties in the evolution history of goldfish.

After studying Chinese historical records, the founder of Chinese goldfish domestication history and genetics research, Mr. Chen Zhen noticed that for 104 years from 1163 to 1267, the breeding way of goldfish had principally been pond.

During this period, only two varieties emerged: white and calico. There were no records about body shape variation. After the bowl feeding came into vogue, for 96 years from 1547 to 1643, it has become a major breeding way. At this stage, as many as 10 new varieties were recorded.

Although people can observe and select goldfish more carefully in this period, the selection was focused on the shape of goldfish. The shape of goldfish has not yet been combined with the selection of fry and parents.

After the late Ming Dynasty and early Qing Dynasty, there were more and more records about goldfish breeding and selection. Bowl feeding enabled people to take a closer look at goldfish and distinguish different varieties from more details. This suggested that people had begun to notice the growing size of goldfish and realized that goldfish had the same shape and color during embryonic development and hatch, but differentiated in shapes and colors later. This was completed during growth and development. We need to select artificially from childhood. While this selection enabled accidental variation to be reserved in the larval stage and completely manifested after they were bred to adult fish.

The core technology of goldfish breeding includes reproducing and raising goldfish and cultivating variety characters. The former is foundation and the latter is core. Goldfish becomes goldfish because different from other fish varieties, it has cultural connotations and aesthetic values. When people begin to realize that the same shape and color in the larval stage will gradually develop into different variety features, selecting larva has become one of the most important links in goldfish feeding. While the selection criteria vary in different areas and ages and are strongly subject to subjective factors. Due to the impact of subjective factors on artificial selection, cultural background and aesthetics in different ages are embodied in goldfish. This is also the source of cultural connotations of goldfish. Until today, the selection criteria and methods of goldfish still vary a lot in different fish farms and become one of the main "secrets" of goldfish cultivation in fish farms.

In the stage of artificial selection, people select fry from the larval stage. Along with the artificial selection of fish, people begin to pay attention to the effect of parents on offspring and select parents consciously for hybridization, according to the desired characters. Hybridization is an important tool for people to transform goldfish and enables them to further create shapes of goldfish based on their understanding of beauty. Mastering hybridization fills the last gap in the goldfish cultivation technology system. From then on, a technical route of goldfish cultivation, i.e., "fry selection- breeding- parent selection–hybridization- artificial reproduction" has been formed and used until today. Once a new character is formed and screened steadily for many times, it will eventually become a new variety.

中国金鱼变异与分类

文 / 曹峰

一、中国金鱼的变异

金鱼是中国人通过长期家化驯养，由野生鲫鱼演化而来。是在外部形态和色彩上，都与其祖先有着较大差异的一种观赏鱼。从金鱼的定义可知，金鱼的变异主要包含外部形态和色彩两个大部分。

1. 外部形态变异

金鱼在外部形态上的变异主要分为鱼鳍的变异、头型的变异、眼型的变异、鼻膜的变异、背峰的变异、鳞片的变异、下颌的变异、鳃盖的变异等几个大类。

鱼鳍的变异

金鱼鱼鳍的变异主要表现为部分品种金鱼的背鳍完全消失，成为蛋形金鱼。臀鳍由单片变为双片，也有少部分返祖为单片臀鳍，其等级要较双臀鳍次一等，而完全失去臀鳍的则作为废品予以淘汰。尾鳍由正常的单尾鳍变为由两片尾叶组成的开尾，根据其外形变化而作为尾鳍变异类的有蝶尾类和孔雀尾类，凤尾、宽尾、裙尾、扇尾、短尾等不作为单独的变异类列出。

头型的变异

金鱼头型的变异是金鱼变异所有类别中最大的一个门类，其主要表现为各种头茸的变异。金鱼的头茸是长期定向选育的结果，其上皮组织退化解体，细胞层次逐渐减少变薄，并被疏松的结缔组织和黏液所替代。从头茸分布位置不同和形状不同，又可以分为鹅头型、虎头型、狮头型和龙头型四大类型。

眼型的变异

金鱼眼型的变异主要表现为眼球外凸于眼眶的龙睛型和望天型，以及下眼眶与眼球之间形成泡囊的蛤蟆眼型和水泡型。眼型变异中另外一种变异是金鱼眼睛的角膜异常发达外凸，如同白炽灯灯泡一般的形状，称为灯泡眼型。灯泡眼型遗传不稳定，且多与其他变异组合形成复合变异。

鼻膜的变异

金鱼鼻膜的变异主要表现为其吻的顶端、鼻腔的两个开口处之间各有一

个小的瓣膜，称为鼻隔，鼻隔先端有肉质
的叶异常发达，呈现球状，所以也叫绒球
或绣球。发育较好的有双球和四球之分，
单球的一般作为次品淘汰。

背峰的变异

金鱼的背峰变异过去鲜有提及，代表
品种有琉金和高身龙睛。这个类型的变异
特点明显有别于其他变异类型，是一个新
的变异类型，它将一般金鱼梭形的侧视体形
变为圆盘形，并将其作为观赏点，属于适合
侧视玩赏的类型。

鳞片的变异

金鱼的鳞片变异主要表现为珍珠鳞，它
是普通鳞片发生钙化沉积后，形成凸起呈小
丘状，附在体表如同一粒粒珍珠。另鳞片中
还有软鳞、透明鳞、荧鳞等不同类型，因其
主要为反光组织结构造成的花色上的不同，
故而归入色彩变异，不作为形态变异描述。

下颌的变异

金鱼的下颌变异主要是由于下颌膜变
异而产生泡囊，其代表品种为各类戏泡金
鱼，由于普及度不高，本书未作详细记述。

鳃盖的变异

金鱼的鳃盖变异主要表现为翻鳃，
是由于金鱼在发育过程中，钙和磷的沉
积不足而造成鳃盖外缘向前翻折，现出
鳃丝。过去翻鳃曾作为一个单独的变异
品种列出，但由于美感度低，较难接
受，因此大多数研究主张将这个变异类
型舍去。目前金鱼业者也大多将翻鳃的
金鱼作为次品淘汰。

2. 色彩变异

金鱼与其祖先鲫鱼的另一大差异就是
产生了五颜六色的色彩变异。金鱼形与色
之辩长久以来广泛存在，到底哪个该优先
也一直在争论。形与色其实是金鱼审美里

一个有机的整体，二者不应被孤立看待。无形，则色无以为附；无色，则形无以所显。

金鱼之所以有五颜六色的变异，是与其真皮层内和鳞片内色素细胞的组成、位置分布以及形状大小决定的。金鱼和鲫鱼一样，主要有3种色素细胞——橙黄色色素细胞、黑色色素细胞和具有淡蓝的分光色的反光组织。由此可见，金鱼颜色的变异与金鱼的鳞片有着紧密的联系。因此，本书将金鱼的色彩变异分为两大类——硬鳞类型和软鳞类型。

硬鳞类型

硬鳞类型是相对于软鳞类型而言的正常鳞片，其色彩包括纯色系、双色系、三色系和多色系。纯色系为单一色彩，有红色、白色、黑色、青色、蓝色、紫色、雪青色。双色系有红白、红黑、黑白、蓝白、紫白、紫红、紫蓝。三色系有红黑白三色和紫红白三色两种。多色系一般指荧鳞色或云锦色。

双色系中的蓝白、黑白、紫红和紫白，三色系中的两种类型都属于不稳定色系。

荧鳞色和云锦色严格意义上属于软鳞硬鳞兼有的属性，但以硬鳞为主的方属于上品。荧鳞侧线以下部分都是硬鳞，只在背脊少量部分为软鳞。云锦色从荧鳞色中选育而出，其硬鳞部分呈现黑色，鳞片边缘呈亮银色或暗金色，头顶和背脊软鳞处为殷红色。

软鳞类型

软鳞类型是由于金鱼不具淡蓝色反光组织，从而呈现透明状态的鳞片，就像玻璃镜没有背面的水银一样。不具色素细胞的软鳞金鱼体表呈现粉白色，有时鳃丝组织和内脏隐约可见。过去对软鳞花色分类不细致，一般只记述软鳞红白和五花，其他花色鲜有记述，而本书将目前流行的各种软鳞花色做了较细致的分类。软鳞花色一般都从软鳞五花中选育而出，呈现双色和多色状态。双色系有软鳞红白（又称樱花）、软鳞红黑（又称虎纹）、软鳞黑白（又分云石和水墨两种）、软鳞紫白、软鳞紫蓝。多色系的称软鳞五花，其中有麒麟、撒锦、重彩、五花等等。

樱花是软鳞红白色，引用日本金鱼同类颜色的命名。基色为如玉般的软鳞白地，殷红色点染其上，面积以少胜多，偶有少量硬质反光鳞点缀，如飘落水面的落樱花瓣。

虎纹是软鳞红黑色，橙红色的底色上着纵向黑色条纹，如同虎皮花纹一般。

云石是软鳞黑白色，白色面积远胜于黑色面积，如同中国山水画里隐约在云海雾涛中的点点山头。

水墨也是软鳞黑白，但黑色面积远胜白色面积，如同中国水墨画，墨色在生宣纸上渗化晕染的效果。

麒麟一般是全身青灰色为基调，白肚，有硬质反光鳞点缀，身上纯青灰色的为墨麒麟，带花斑的称花麒麟。头部以红顶为贵，玉顶次之。

撒锦又称芝麻花，一般以素蓝色或米白色为地，上面点缀细碎黑点，偶有少量硬质反光鳞点缀。

重彩是指体表被大面积墨色和红色覆盖，白色和素蓝色底色面积很少，如同大泼墨泼彩的中国画一般。

五花是传统软鳞花色的经典代表，它要求素蓝色的背，配以白腹红顶，点缀少量黑色和橙黄色的斑点。五花色暗合中国的五行之色，但高品质的五花色极少。

二、中国金鱼的分类

中国金鱼的分类和生物学上的系统分类是不同的概念，它更多是以金鱼的外部形态特征变异和遗传特征作为分类标准，从而进行归类总结。而一直以来较多采用的是源自《竹叶亭杂记》中所记载的草种、文种、龙种、蛋种的四分法。也有草、文、蛋的三分法，将龙背列入的五分法，甚至还有很细的十三种类法。三分法的问题首先在于将草种单独列出意义不大，该系统内没有足够的品种数量支撑这个系统；其次，相对于蛋种来说，它和文种的界定不十分明显。四分法除上述问题外，龙种的界定又采用另外的判断标准，使同一个层面上的4个平行分类，却各自使用3个不同的判定标准，比较混乱。因此，本书在国内各地专家讨论的基础上，采用二分法对金鱼进行分类，即按照外形上有无背鳍这一唯一标准，将中国金鱼划分为文形和蛋形两大系统，并以文形金鱼和蛋形金鱼分别作为两个系统的基本型，在基本型的基础上叠加其他各种变异，从而涵盖金鱼的各种特征变异类型。

金鱼的品种分类划分并不代表金鱼演化的先后次序，它仅仅是将目前已知金鱼变异的状况按照一定的规律进行归纳和总结。文形和蛋形两大金鱼系统不是相互孤立、独自发展的，期间或有交叉叠合的现象，如有的蛋形金鱼品种是由文形金鱼演化而来的，而有的文形金鱼品种是由蛋形金鱼演化而来的。

中国金鱼的命名与其变异密不可分，一般就是"颜色+变异"，例如红白水泡。但长期以来，对变异类型（主要是头型变异）的定义，有着不同的称谓，因此造成金鱼命名的困扰与纷争。本书认为，虽然许多名称现在看是有讹误，但已经约定成俗，不宜轻易否定。我们在尊重传统金鱼命名的基础上，以注释的形式对其变异特征加以说明，称为二段式命名。例如红白寿星（狮头型），就是表示传统金鱼的红白寿星，其头型变异为狮头型。

Chinese Goldfish Variation an Classification

Article / Cao Feng

1 Chinese Goldfish Variation

Goldfish has been long domesticated evolving from wild carp. It is distinctive form their ancestor on shape and color. From the definition of goldfish, the main variations are focused on shape and color.

1.1 Morphology Variation

The main variation of shape is mainly on fins, head, eyes, nasal valve, back hump, scales mandible and gills.

Fin Variation

Fin variation happens on dorsal fins, anal fins or caudal fins. If the dorsal fins disappear, it changes to egg shape; sometimes, anal fin changes from single to pair, few remains single, but is in lower grade than the latter one; those completely lost the anal fin should be weeded out; the caudal fin changes from single to pair, exemplified by butterfly tail type or cock tail type, but phoenix tail, wide tail, dress tail, fan tail, or short tail do not fall into the same category.

Head Variation

Head variation is the biggest category among all the variations. The major change is on

the head pompon. After oriented selective breeding, the epithelial tissue faded, replaced by loosen connective tissue and mucus, make the cell layers thinner. According to head pompon position and shape, the variations are classified as Goosehead, Tigerhead, Lionhead and Dragonhead.

Eye Variation

Eye variation refers to Dragon Eyes and Celestial Eye with eyes protruding outside the obit, or Frog Eye or Bubble-Eye with vesicle between lower obit and eyeball. Another eye mutants is what we call bulb eye with its extraordinarily developed cornea pointed out. It's not stable genetic, so normally as part of the compound variation.

Nasal Variation

Nasal variation happens on the upper end of snout. There is a small septum between two nasal holes, called nasal septum. It is ball shaped, very developed, so it is also called pompon. Two-pompon and four-pompon are most popular, single-pompon is regarded defective, and will be weed out.

Back Hump Variation

It is seldom mentioned about the height hump variation of goldfish. With the appearance of varieties of Ryukin and high hump dragon eyes, apparently different from other variation, the height variation is listed out as a new category. Height variation is better to side-view appreciation for its round plate shape.

Scale Variation

The scale variation refers to pearl scale, which is mutated from ordinary scale after sediment of calcium, protruding like pearl. Other variation like Soft-Scale, transparent scale, florescence scale and so on, are characterized as color mutants instead of shape mutants due to it is caused by their reflective cells structure.

Mandible Variation

Mandible variation refers to variation with vesicle beneath the mandible. Representative varieties are all kinds of bubble goldfish. They are not well introduced due to less prevalent and restriction of this book.

Gill Variation

Gill variation refers to reverse of the gill, caused by the sediment of calcium and phosphorus during growth. It was categorized as a unique mutant before, but was eliminated due to lack of beauty. Now they are also weed out by many raisers.

1.2 Color Variation

The main difference between goldfish and their ancestor wild carp, lies in the color variations. The discussion about color and shape as for which one comes first lasts for quite a long time. Color and shape should not be isolated, they are an entire integrity. No shape, color cannot attach, no color, shape will be dull and boring.

The color variation is decided by the pigment cells of dermis and scales with different locations and shapes. Both goldfish and carp have these three dermis cells: orange dermis, black dermis and the reflective blue tissues. Therefore, the variation of color has an indispensable relation with scales of the goldfish. In this book, the color variation is divided into two major types: Hard-Scale type and Soft-Scale type.

Hard-Scale Type

Hard-Scale type is much normal compared with Soft-Scale type. It is consisted of single-color, bi-color, tri-color and multi-color. Single-color includes red, white, black, green, blue, chocolate, and lilac. Bi-color includes red white, red black, black white, blue white, chocolate white, chocolate red, chocolate blue. Tri-color includes red black white and chocolate red white. While multi-color refers to florescence scale or Yunjin color (Yunjin, gold thread used for emperor's robe).

Bi-color of blue white, black white, chocolate red and chocolate white, as well as the tri-colors type are both not stable.

Florescence scale and Yunjin scale, in strict sense, both have Hard-Scale variety and Soft-Scale variety, but the former one is more preferred. Beneath the side line is the Hard-Scale, only a few of Soft-Scale on the back area. Yunjin type is selective bred from florescence scale type, black on hard area, bright silver or dark golden on the scale edge, red on the head and dorsal fin.

Soft-Scale Type

Soft-Scale type is like the transparent mercury of a mirror without the coating, since it does not have blue reflective tissues. This variety is in pink transparent, so we can even see the inside part of the gill or the body. In the past, there was no detailed classification on Soft-Scale, only red white scale and calico were recorded. As a supplement, this book has made more subtle descriptions on the classification. Normally, Soft-Scale is developed from Soft-Scale Calico (Bluish Base), presenting bi-color or multi-color. Bi-color is Soft-Scale red white, also called sakura; Soft-Scale red black, or tiger streak; Soft-Scale black white includes marble color and inky color; Soft-Scale chocolate white; Soft-Scale chocolate blue. Multi-color type is Soft-Scale Calico (Bluish Base) Goldfish, consisting of Kirin, Oranda (with Bluish Matt and Black Dots), Intense Color and Calico (Bluish Base) and so on.

Sakura also Soft-Scale red white, is named after the same variety from Japan. It is white and soft, with blackish red spots on it, the less the better, sometimes ornamented by reflective Hard-Scale, like the sakura falling on water.
Tiger streak is red black Soft-Scale. Horizontal black streaks lie on the orange background, looking like the tiger streaks.

Marble refers to the Soft-Scale black white, the white should be larger than the black, just like the mountains on the ink paintings.

Water pattern, Soft-Scale black white, with black oversized the white, as vivid as a traditional water ink paintings.

Kirin, green and grey in general, white belly with reflective Hard-Scale. Pure colored called black Kirin, calico colored called flower Kirin. Red pompon is most rare, while jade pompon is ranking after that.
Oranda (with bluish matt and black dots), also called fried dough twist, light

blue or creamy white as the background, cover with tiny black spot, some is with ornamented with reflective Hard-Scale.

Intense color means a large area of black color or red color, less white and light blue, like the inky Chinese painting.

Calico with bluish base is the representative type of Soft Scale. Blue on the back, white on the belly, covered with some black or orange spots. Calico, or "five color", reflect the colors of Five Elements of nature. It's very rare to see those with high quality.

2 The Classification of Chinese Goldfish

Differentiated from biology, the goldfish classification is in another system. It has collected and summarized the genetic characteristics of color and shape of different varieties. Derived from the book *Notes in Bamboo Pavilion*, there are four main categories: Common Goldfish, Fantail Goldfish, Dragon type, Egg-Fish , or three main categories since sometimes Dragon type can be emerged into other types. Moreover, there are even five or thirteen categories. There is much significance for three way classification, which intends to highlight Fantail Goldfish, because first, the number of the species is not considerable, and second, compared with Egg-Fish, it is not well distinguished with Fantail Goldfish. For four-way classification, except above problem, Dragon type is categorized as a new type, at the same time three-way classification is also adopted. After consulting experts throughout the country, this book tries to establish two-way classification that is Fantail Goldfish and Egg-Fish, with dorsal fins or without. Based on these two major types, many other variations are developed.

The classification of goldfish does not reflect the evolution sequence. It is only a summary of goldfish variation per some kind of discipline. Fantail Goldfish and Egg-Fish are independent with each other, they are overlapped sometimes. For example, Egg-Fish is evolved from fantail type, and all the Fantail Goldfish is also evolved from Egg-Fish.

The naming of Chinese goldfish is highly related to its variation, normally is color+ mutant, like Red White Water Bubble. However, the naming definition for variation especially for head variation has not been built up, which has brought many confusions and conflicts. The writers of this book hold the view that, although the naming is not absolutely right, but since they have accepted by people for long time. We should respect the traditional way of naming goldfish, meanwhile to interpret the details of the variations, i.e. supplementary naming. For example, Red White Lionhead (Southern) actually refers to traditional Red White Lionhead (Southern), it belongs to head variation- lion head..

① 背鳍/Dorsal Fin

② 尾柄/Caudal Body

③ 尾芯/Caudal Center

④ 尾鳍/Caudal Fin

⑤ 尾先/Tail Edge

⑥ 亲骨/Tail Bone

⑦ 鳍棘/Fin Spine

⑧ 背/Back

⑨ 侧线/Lateral Line

⑩ 顶茸/Head Pompon

⑪ 眼/Eye

⑫ 鼻膜/Nasal Valve

⑬ 吻凸/Snout

⑭ 口/Mouth

⑮ 鬓茸/Sideburns Pompon

⑯ 颌茸/Jaw Pompon

⑰ 鳃茸/Gill Pompon

⑱ 胸鳍/Pectoral Fin

⑲ 泄殖孔/Cloacal Pore

⑳ 臀鳍/Anal Fin

㉑ 腹鳍/Pelvic Fin

㉒ 尾肩/Middle Tail

中国金鱼变异分类模式图
Mode of Variation of Chinese Goldfish

01. 头型变异类 / Head Variation

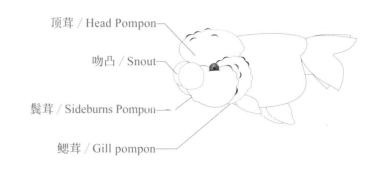

顶茸 / Head Pompon
吻凸 / Snout
鬓茸 / Sideburns Pompon
鳃茸 / Gill pompon

■ 发达 / Well Developed
■ 不发达 / Less Developed

鹅头型 / Goosehead

顶茸（发达）/ Head Pompon (well developed) •
鳃茸（无）/ Gill Pompon (none) •
吻凸（无）/ Snout (none) •
鬓茸（无）/ Sideburns Pompon (none) •
典型品种 (鹅头红、鹤顶红) •
Classical Type (Red Cap Goosehead,Red Cap Oranda)

虎头型 / Tigerhead

顶茸（发达）/ Head Pompon (well developed) •
鳃茸（不发达）/ Gill Pompon (less developed) •
吻凸（不发达）/ Snout (less developed) •
鬓茸（发达）/ Sideburns Pompon (well developed) •
典型品种（红虎头、国狮）/ Classical Type [Red Lionhead (Egg-fish),Oranda]

狮头型 / Lionhead

顶茸（发达）/ Head Pompon (well developed) •
鳃茸（发达）/ Gill Pompon (well developed) •
吻凸（不发达）/ Snout (less developed) •
鬓茸（发达）/ Sideburns Pompon (well developed) •
典型品种（国寿、红狮头）/ Classical Type (Ranchu,Red oranda) •

龙头型 / Dragonhead

- 顶茸（不发达）/ Head Pompon (less developed)
- 鳃茸（不发达）/ Gill Pompon (less developed)
- 吻凸（发达）/ Snout (well developed)
- 鬓茸（不发达）/ Sideburns Pompon (less developed)
- 典型品种（日寿、东锦）/ Classical Type (Japanese Ranchu,Azumanishiki)

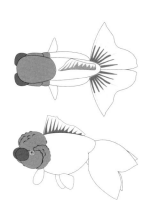

02. 眼型变异类 / Eye Variation

龙睛型 / Dragon Eyes

- 眼型（外凸）/ Eye Shape (evagination)
- 型体特征（眼左右朝向平行）/ Body Feature (parallel eye direction)
- 典型品种（龙睛、蛋龙）/ Classical Type (Telescope,Egg-fish with Dragon Eyes)

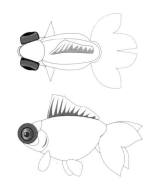

蛤蟆眼型 / Frog-Head

- 眼型（眼下长泡）/ Eye Shape (vesicle beneath eye)
- 型体特征（泡眼小而硬）/ Body Feature (small hard bubble eye)
- 典型品种（蛤蟆眼）/ Classical Type [Frog-Head (Egg-fish)]

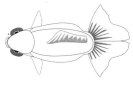

望天型 / Celestial-Eye

- 眼型（外凸上翻）/ Eye Shape (convex upwards)
- 型体特征（双眼朝上）/ Body Feature (eyes grow upwards)
- 典型品种（望天）/ Classical Type (Celestial-Eye)

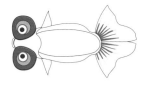
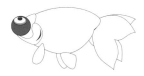

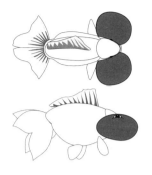

水泡型 / Bubble-Eye

眼型（眼下长泡）/ Eye Shape (vesicle beneath eyes) •
型体特征（泡眼大而软）/ Eye shape (large soft bubble eye) •
典型品种（水泡眼）/ Classical Type (Bubble-Eye) •

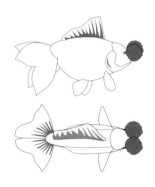

03. 鼻膜变异类 / Nasal Variation

绒球型 / Pompons

鼻瓣膜（发达）/ Nasal Valve (well developed) •
典型品种（文球、蛋绒球）
Classical Type (Fantail Goldfish with pompons,Egg-Fish with pompons)

04. 背峰变异类 / Hump Variation

琉金型 / Ryukin

背峰（背峰高耸）/ Back Hump (high) •
型体特征（侧视圆盘形）/ Body Feature (top view disc) •
典型品种（琉金、高身龙睛）
Classical Type (Ryukin,Telescope with High Hump)

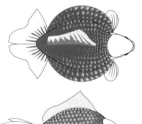

05. 鳞片变异类 / Scale Variation

珍珠鳞型 / Pearlscale

鳞片（凸起）/ Scale (evagination) •
型体特征（身形圆润）/ Body Feature (round and fat) •
典型品种（皮球珍珠）/ Classical Type (Golfball Pearlscale) •

06. 尾鳍变异类 / Caudal Fin Variation

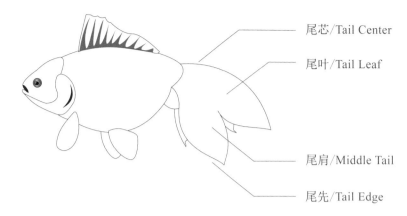

尾芯/Tail Center

尾叶/Tail Leaf

尾肩/Middle Tail

尾先/Tail Edge

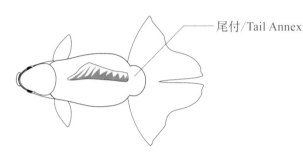

尾付/Tail Annex

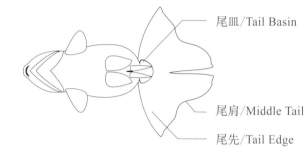

尾皿/Tail Basin

尾肩/Middle Tail

尾先/Tail Edge

蝶尾型 / Butterfly Tail

- 尾芯角度（平直）/ Caudal Position (flat)
- 尾肩（前倾）/ Middle Tail (forward)
- 尾叶（四）/ Tail Leaf (four)
- 典型品种（文蝶）/ Classical Type (Butterfly Tail)

孔雀尾型 / Peacock Tail

- 尾芯角度（上扬）/ Tail Center Position (upward)
- 尾肩（左右平伸）/ Middle Tail (stretching to left and right)
- 尾叶（四）/ Tail Leaf (four)
- 典型品种（地金）/ Classical Type (Jikin)

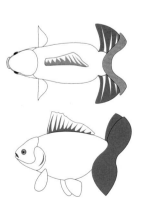

中国金鱼产区分布示意图
The Origins Indicator for Chinese Goldfish

新兴产地 / Rising Production Places

G. 湖南 / Hunan	**K.** 湖北 / Hubei	**O.** 重庆 / Chongqing
H. 河北 / Hebei	**L.** 山东 / Shandong	**P.** 甘肃 / Gansu
I. 安徽 / Anhui	**M.** 云南 / Yunnan	**Q.** 台湾 / Taiwan
J. 辽宁 / Liaoning	**N.** 陕西 / Shanxi	

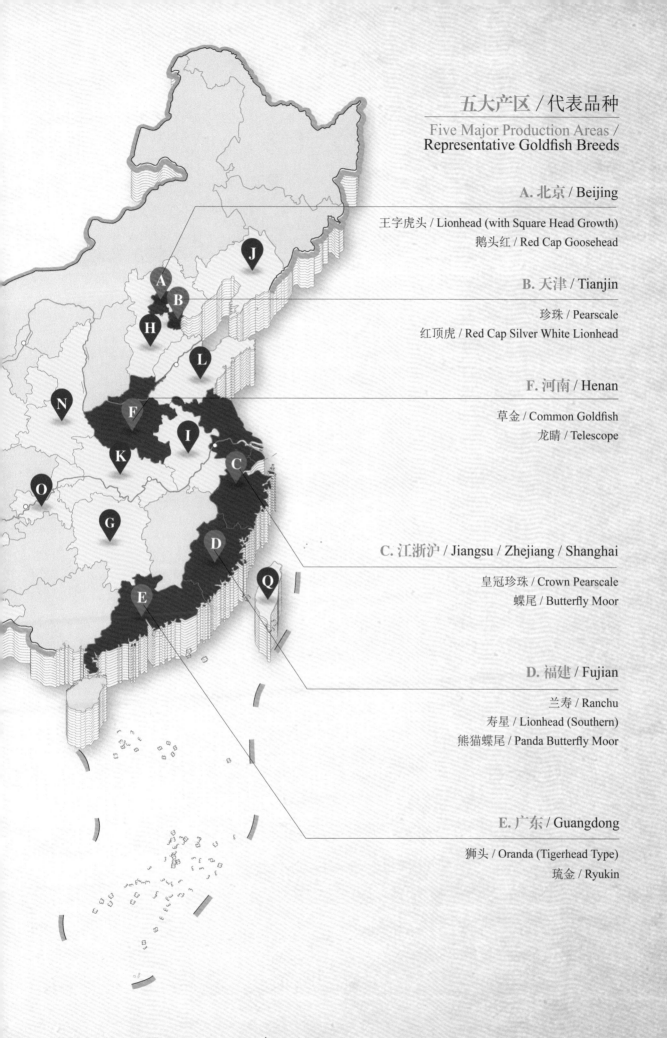

五大产区 / 代表品种
Five Major Production Areas /
Representative Goldfish Breeds

A. 北京 / Beijing

王字虎头 / Lionhead (with Square Head Growth)
鹅头红 / Red Cap Goosehead

B. 天津 / Tianjin

珍珠 / Pearscale
红顶虎 / Red Cap Silver White Lionhead

F. 河南 / Henan

草金 / Common Goldfish
龙睛 / Telescope

C. 江浙沪 / Jiangsu / Zhejiang / Shanghai

皇冠珍珠 / Crown Pearscale
蝶尾 / Butterfly Moor

D. 福建 / Fujian

兰寿 / Ranchu
寿星 / Lionhead (Southern)
熊猫蝶尾 / Panda Butterfly Moor

E. 广东 / Guangdong

狮头 / Oranda (Tigerhead Type)
琉金 / Ryukin

本书使用说明
Introduction of This Book

本书突破了中国金鱼的传统分类法，采用最新分类法"二分法"，对我国金鱼品种的分类重新进行了系统的梳理。

本书引入科普图解——中国金鱼变异分类模式图，读者可以直观地了解到金鱼不同变异类型的形态特征。开篇部分附有中国金鱼产区分布示意图，展示中国金鱼的传统产区与新兴产区的分布现状。本书条目以图文并茂的形式，标明了每种金鱼的分类信息，读者不仅可以查阅具体品种的形态特征、习性、历史等信息，还可以比照精选图片和鉴赏文字对金鱼进行辨识赏析。本书还特别提供了中英文对照，为海外读者了解中国金鱼开辟了新渠道。

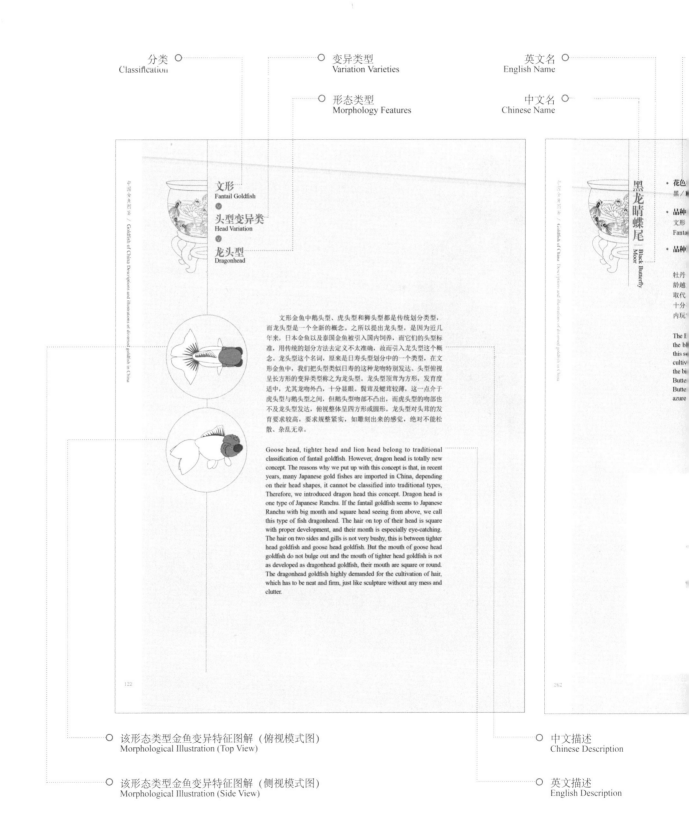

分类
Classification

变异类型
Variation Varieties

英文名
English Name

形态类型
Morphology Features

中文名
Chinese Name

中国金鱼图鉴 / Goldfish of China Descriptions and illustrations of diversed goldfish in China

文形
Fantail Goldfish
∨
头型变异类
Head Variation
∨
龙头型
Dragonhead

文形金鱼中鹅头型、虎头型和狮头型都是传统划分类型，而龙头型是一个全新的概念。之所以提出龙头型，是因为近几年来，日本金鱼以及泰国金鱼被引入国内饲养，而它们的头型标准，用传统的划分方法去定义不太准确，故而引入龙头型这个概念。龙头型这个名词，原来是日寿头型划分中的一个类型，在文形金鱼中，我们把头型类似日寿的这种龙吻特别发达、头型俯视呈长方形的变异类型称之为龙头型。龙头型顶茸为方形，发育度适中，尤其龙吻外凸，十分显眼。鬓茸及鳃茸较薄，这一点介于虎头型与鹅头型之间，但鹅头型吻部不凸出，而虎头型的吻部也不及龙头型发达，俯视整体呈四方形或圆形。龙头型对头茸的发育要求较高，要求规整紧实，如雕刻出来的感觉，绝对不能松散、杂乱无章。

Goose head, tighter head and lion head belong to traditional classification of fantail goldfish. However, dragon head is totally new concept. The reasons why we put up with this concept is that, in recent years, many Japanese gold fishes are imported in China, depending on their head shapes, it cannot be classified into traditional types, Therefore, we introduced dragon head this concept. Dragon head is one type of Japanese Ranchu. If the fantail goldfish seems to Japanese Ranchu with big mouth and square head seeing from above, we call this type of fish dragonhead. The hair on top of their head is square with proper development, and their mouth is especially eye-catching. The hair on two sides and gills is not very bushy, this is between tighter head goldfish and goose head goldfish. But the mouth of goose head goldfish do not bulge out and the mouth of tighter head goldfish is not as developed as dragonhead goldfish, their mouth are square or round. The dragonhead goldfish highly demanded for the cultivation of hair, which has to be neat and firm, just like sculpture without any mess and clutter.

122

中国金鱼图鉴 / Goldfish of China Descriptions and illustrations of diversed goldfish in China

黑龙睛蝶尾
Black Butterfly Moor

· 花色
黑 /

· 品种
文形
Fanta

· 品种

牡丹
龄越
取代
十分
内玩

The I
the bl
this se
cultiv
the bi
Butter
Butter
azure

262

该形态类型金鱼变异特征图解（俯视模式图）
Morphological Illustration (Top View)

中文描述
Chinese Description

该形态类型金鱼变异特征图解（侧视模式图）
Morphological Illustration (Side View)

英文描述
English Description

This book not restricted to the traditional goldfish classification, has adopted dichotomy to reorganize the existing varieties of Chinese goldfish.

With the popular science illustration, readers can well understand different shapes and characteristics of different variations. It starts with a map of goldfish origins, traditional and newly risen. This book has used many legends to illustrate different categories and its classification. Not only can readers understand the features, habits, history and origins, but also have a nice appreciation on the goldfish pictures and depictions. As it is also a book for foreign readers, this book has translated into English at the same time.

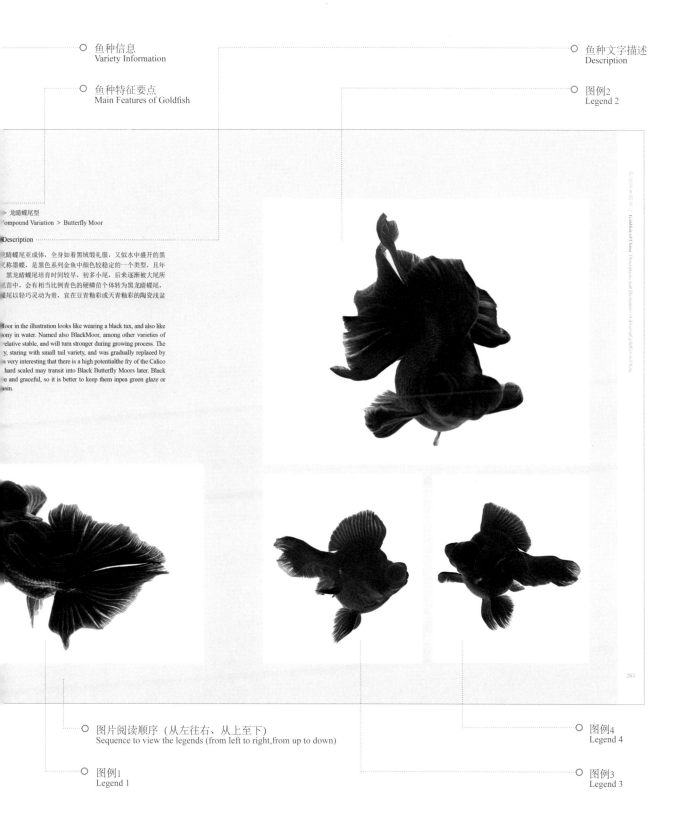

○ 鱼种信息
Variety Information

○ 鱼种特征要点
Main Features of Goldfish

○ 鱼种文字描述
Description

○ 图例2
Legend 2

> 龙睛蝶尾型
ompound Variation > Butterfly Moor

Description

龙睛蝶尾亚成体，全身如着黑绒缎礼服，又似水中盛开的黑
又称墨蝶，是黑色系列金鱼中颜色较稳定的一个类型，且年
黑龙睛蝶尾培育时间较早，初多小尾，后来逐渐被大尾所
尾中，会有相当比例青色的硬鳞苗个体转为黑龙睛蝶尾，
尾以轻巧灵动为贵，宜在豆青釉彩或天青釉彩的陶瓷浅盆

Moor in the illustration looks like wearing a black tux, and also like
ony in water. Named also BlackMoor, among other varieties of
relative stable, and will turn stronger during growing process. The
, staring with small tail variety, and was gradually replaced by
s very interesting that there is a high potentialthe fry of the Calico
hard scaled may transit into Black Butterfly Moors later. Black
e and graceful, so it is better to keep them inpea green glaze or
asin.

中国金鱼图鉴 / Goldfish of China: Description and illustration of diversed goldfish in China

263

○ 图片阅读顺序（从左往右、从上至下）
Sequence to view the legends (from left to right,from up to down)

○ 图例4
Legend 4

○ 图例1
Legend 1

○ 图例3
Legend 3

文形
Fantail Goldfish

中国金鱼从外部形态上可以区分为两大类别，即文形和蛋形，这也可以作为中国金鱼分类树的两大基本分支。文形金鱼的名称由其原型类品种文鱼而来，它和蛋形金鱼的唯一区别就是有无背鳍，具有完整背鳍的金鱼全部归为文形金鱼。

文鱼直接从原始红色鲫鱼演化而来。"文"字的本义是指花纹、纹理。汉代许慎《说文解字》中说："文，错画也，象交文。"意思是说，"文"字是指"交错画的花纹"。因此，古人将有颜色和花纹的鱼统称为文鱼，例如屈原《楚辞•九歌•河伯》篇中有"乘白鼋兮逐文鱼，与女游兮河之渚"之句。而当金鱼逐渐成为中国观赏鱼的主要类别之后，文鱼便专指由红鲫鱼进化而来的、具有开尾和色彩变化的原始金鱼。

文形金鱼出现早于蛋形金鱼，其覆盖的品种数也略多于蛋形金鱼。以文形金鱼为纲，其下可划分出原型类、头型变异类、眼型变异类、鼻膜变异类、背峰变异类、鳞片变异类、尾鳍变异类、复合变异类和多重复合变异类。

Chinese goldfish can be classified into two types according to their shapes, i.e., Fantail Goldfish and Egg-Fish, which are also the two basic categories of Chinese goldfish. The name of Fantail Goldfish is originated from its Original Type fantail which distinguishes itself from Egg-Fish by possessing dorsal fin. Any goldfish which has a full featured dorsal fin is classified into Fantail Goldfish.

Fantail Goldfish is an evolution of ancient red carp. The original meaning of "wen"is pattern and texture. As per *Shuo Wen Jie Zi* wrote by Xu Shen in Song Dynasty, "wen" was interpreted as "staggered pattern". Therefore, the ancients named all fishes with color and pattern as wen fish. For example, in one of the poems of an ancient poet Qu Yuan, there is a sentence: "Riding white trionyx sinensis to chase wen fish, and wandering the eyot with you". As the goldfish becomes the main aquarium fish in China, Fantail Goldfish has been specifically referring to ancient goldfish which was evolved from red carp and bearing the features of split tail and variation of color.

Fantail Goldfish appeared earlier than Egg-Fish, and has more varieties than Egg-Fish. Within the category of Fantail Goldfish, there are original type, types with variations at head, eye, nose, hump, dorsal fin, scale and caudal fin, as well as types with compound variations.

文形
Fantail Goldfish

∨

原型类
Original Type

∨

草金型
Common Goldfish

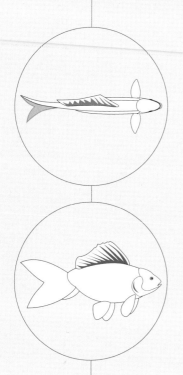

文形金鱼的原型类涵盖各种颜色的花色鲫鱼和文鱼，花色鲫鱼又称为草金，其体形与普通鲫鱼基本没有区别，只是鱼鳍略有变长，并且多了色彩的变异。原始的草金颜色变化并不多，一般为红色、橙色、白色和红白色。而目前草金多彩的颜色变化，多为和现代金鱼杂交选育而出。

花色鲫鱼体形介于金鱼和锦鲤之间，群聚性和游动性强，比较适合庭院中池塘养殖和观赏，也可与锦鲤搭配混养，是现代园林水族造景的重要素材之一。

The original type of Fantail Goldfish includes colored carp and those with various colors. Colored carp, also called Common Goldfish, has same size with normal carp, but longer fin and varying color. Original Common Goldfish has dull color, in red, orange, white or red white. The rich color variation of Common Goldfish at present is because of hybridization with modern goldfish.

The size of colored carp is appropriate, larger than goldfish while smaller than Koi. Due to gathering habitual nature and mobility, it's suitable for being bred in courtyard pools with fancy carp, thus forming one of the important elements of aquascape for modern garden.

红草金
Red Common
Goldfish

花色类别 / Color •
红 / Red

品种类别 / Species •
文形 ＞ 原型类 ＞ 草金型
Fantail Goldfish ＞ Original Type ＞ Common Goldfish

品种详述 / Species Description •

　　红草金又称金鲫鱼或红鲫鱼，体形与普通家鲫鱼基本近似，其尾鳍稍长。原始的金鲫鱼一般为橙红色，而随着不断进行人工选育，现代的红草金全身呈火红色。红草金野性强，适宜群聚，可以较大水面养殖，耐粗放管理。

Red Common Goldfish, also named Golden Carp or Red Carp, has similar size and longer caudal fin as compared to normal carp. Original golden carp is generally orange red. Along with continuous artificial selection, modern Red Common Goldfish is red like a fire. Red Common Goldfish, with inborn wild and gathering nature, is suitable for breeding in large water and extensive management.

- **花色类别** / Color

  黑 / Black

- **品种类别** / Species

  文形 > 原型类 > 草金型

  Fantail Goldfish > Original Type > Common Goldfish

- **品种详述** / Species Description

  黑草金又称墨色草金，外部体形特征同红草金，全身黑色，腹部多为银灰色或橙黄色。红草金中有个别个体会在褪色期变为黑色，但不稳定，后期会逐渐变为红色。而五花草金中的青色个体也有部分会转为黑色且较稳定。

  Black Common Goldfish, or Inky Common Goldfish, shares appearance characteristics with Red Common Goldfish. Black colored in general, while silver gray or orange at the belly. Some Red Common Goldfish may turn to black during certain period, then resume red gradually later. Some Calico Common Goldfish (Bluish Base) may turn to black and then stabilized.

花色类别／Color •

白／White

品种类别／Species •

文形 ＞ 原型类 ＞ 草金型

Fantail Goldfish ＞ Original Type ＞ Common Goldfish

品种详述／Species Description •

白草金
White Common
Goldfish

　　白草金是红白草金中褪白的个体，一般幼鱼时期即被淘汰。在群体中做搭配色彩时会留极少个体，或在繁育时使用。

White Common Goldfish refers to Red White Common Goldfish which turn white. In general, such fish will be weeded out during their juvenile period. Few White Common Goldfish will be kept in the group for enriching color, or used for breeding.

白草金

紫草金

Chocolate Common Goldfish

- **花色类别** / Color
  紫 / Chocolate

- **品种类别** / Species
  文形 > 原型类 > 草金型
  Fantail Goldfish > Original Type > Common Goldfish

- **品种详述** / Species Description

  紫草金与金鱼中的紫色个体相近，日本将金鱼中这个色系称为茶色，其色彩不稳定，较易在成长后期转色为红草金，在混养中作为配色使用，其他形态特征和红草金完全相同。

  Chocolate Common Goldfish is similar with the chocolate goldfish. In Japanese goldfish, such color is also called tawny. Its color is not stable, and easily to turn to Red Common Goldfish after growing up. Chocolate Common Goldfish is used for enriching color in mixing breeding. Its morphological characteristics are the same as Red Common Goldfish.

花色类别 ／ Color •

红白 ／ Red White

品种类别 ／ Species •

文形 ＞ 原型类 ＞ 草金型

Fantail Goldfish ＞ Original Type ＞ Common Goldfish

品种详述 ／ Species Description •

红白草

金

Red White

Common Goldfish

　　红白草金和五花草金都是草金中最受市场欢迎的品种。由于体形较金鱼个体要大，所以可以很好地表现出红白花色的纹路变化。由于受日本红白锦鲤的影响，这几年红白草金的花色玩赏，有向红白锦鲤借鉴的趋势。红白草金适宜较宽水域饲养，宜群聚，观赏性较强。

Red White Common Goldfish and Calico Common Goldfish (Bluish Base) are popular types of Common Goldfish in the market. As its size is larger than goldfish, the details of red and white pattern can be displayed better. As influenced by the Red White Koi from Japan, the beauty-appreciation for Red White Common Goldfish is in line with the Red White Koi. Red White Common Goldfish is suitable for large water, with gathering nature and high ornamental value.

红黑草金
—— Red Black Common Goldfish

- **花色类别** / Color
  红黑 / Red Black

- **品种类别** / Species
  文形 > 原型类 > 草金型
  Fantail Goldfish > Original Type > Common Goldfish

- **品种详述** / Species Description

  红黑草金多为黑草金向红草金转色的一个过渡期品种。其他特征与红草金完全相同。

  Red Black Common Goldfish is a transition variety from Black Common Goldfish to Red Common Goldfish. Its characteristics are totally the same as Red Common Goldfish.

花色类别 / Color •

软鳞红白 / Red White Matt

品种类别 / Species •

文形 > 原型类 > 草金型

Fantail Goldfish > Original Type > Common Goldfish

品种详述 / Species Description •

樱花草金 | Sakura Common Goldfish (Red White Matt)

樱花草金是五花草金中失去蓝色的变异个体，一般呈现出软鳞红白花色，因红色绯块上点缀的点点白色硬鳞如同水中的落樱，故而得名樱花。樱花草金以白色面积多于红色面积为佳，樱花鳞宜少不宜多，所谓点到即止。其他形态特征与五花草金完全相同。

Sakura Common Goldfish (Red White Matt) is a variation of Calico Common Goldfish (Bluish Base) which don't have blue color. In general, it's in red white matt. Its name is inspired by the sakura-like pattern formed by red area decorated white Hard-Scale. The white area shall be larger than the red area. Few sakura scale is enough. Its morphological characteristics are the same as Calico Common Goldfish (Bluish Base).

花草金 | Calico Common Goldfish

- **花色类别** / Color
  软鳞三色 / Calico

- **品种类别** / Species
  文形 ＞ 原型类 ＞ 草金型
  Fantail Goldfish ＞ Original Type ＞ Common Goldfish

- **品种详述** / Species Description

　　花草金是五花草金中失去一两种色调的软鳞花色草金，多呈现出软鳞三色红白黑，其中以红白为主，点缀少许黑色斑点或黑色条纹即可，若黑色面积过大则易显脏，尾鳍处分布有黑色条纹则品相佳，其他形态特征与五花草金完全相同。

Compared with Calico Common Goldfish (Bluish Base), Calico Common Goldfish has fewer colors, and its scale is generally in red, white and black. Red and white are the main colors, trimmed with few black spot or strip. It may make people feel dirty if the black area is too large. Black strips at the caudal fin make outstanding appearance. Except the aforementioned, it shares same morphological characteristics with Calico Common Goldfish (Bluish Base).

花色类别 / Color •

软鳞五花 / Calico (Bluish Base)

品种类别 / Species •

文形 > 原型类 > 草金型
Fantail Goldfish > Original Type > Common Goldfish

品种详述 / Species Description •

五花草金
Calico Common
Goldfish (Bluish Base)

　　五花草金又称花草，其中长尾型的又称燕尾花草，与红草金不同，它是通过五花色金鱼选育而出，并非原始花色。五花草金与其他花色草金或锦鲤搭配饲养，可以取得较好的观赏效果。

Calico Common Goldfish (Bluish Base), either called Calico Common Goldfish, if with long tail, also called Fork Tail Common Goldfish. Unlike Red Common Goldfish, it is developed from goldfish with various colors, not born color. Calico Common Goldfish (Bluish Base) along with Carp in other colors or Koi can enable a good combination of aquarium.

软鳞五花 / Calico (Bluish Base)

文形
Fantail Goldfish

⋁

原型类
Original Type

⋁

文鱼型
Fantail Goldfish

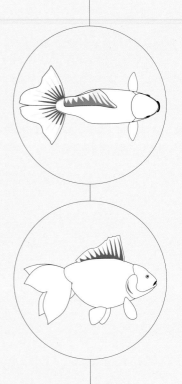

文鱼身体形态上相较于普通鲫鱼的长梭形，更趋向短宽，具有较长的鱼鳍，尾鳍为四开尾，无其他特征变异。文鱼是位于金鱼进化树起始端的一个原始品种，在客观现实中，由于金鱼进化的持续向前，纯正意义上的文鱼早已湮没在时空的长轴之中，而我们现在看到的文鱼，多为通过杂交手段筛选出的一个近似文鱼的品种类别，也可称之为现代文鱼。文形金鱼中的琉金，是由文鱼直接演化过去的，因此，在琉金中还会出现一些返祖个体，比如体高收窄，它是比较接近文鱼原始形态的。由于文鱼是比较原始的一个品种，其观赏性被其他品种所取代，多在品种杂交选育中作为一个载体使用。

Compared with normal carp which is fusiform spindle-shaped, Fantail Goldfish is more short and wide, with long fins and split caudal fin. It is the most original variety during the evolution of goldfish. In fact, due to the constant evolution, pure Fantail Goldfish has already extinct in the long history. The new Fantail Goldfish breed at present is a variety similar to Fantail Goldfish by hybridization. We can call it modern Fantail Goldfish. Ryukin is a direct offspring of Fantail Goldfish. Therefore, among Ryukin, there may be some reversion individuals which is long and narrow, similar to the original morphology of Fantail Goldfish. As it is an old variety, many other varieties have replaced its place as ornamental fish. It is generally used as a carrier for hybridization.

花色类别 / Color •

白 / White

品种类别 / Species •

文形 > 原型类 > 文鱼型

Fantail Goldfish > Original Type > Fantail Goldfish

品种详述 / Species Description •

　　图例所示为白文鱼。白文鱼身体为梭形，具四开尾，眼睛、头部、鳞片等正常无变异。二龄以后尾鳍变长，甚至可以达到体长的两倍，有一定的观赏性。体质强，较易饲养。

The figure shows White Fantail Goldfish which has fusiform spindle-shaped body, split caudal fin, and normal eye, head and scale. Its caudal fin grows two years later, some even two times long as the body, which features certain ornamental value. Robust and easy to feed.

白文鱼
White Fantail Goldfish

蓝文鱼——Blue Fantail Goldfish

- **花色类别** / Color

  蓝 / Blue

- **品种类别** / Species

  文形 > 原型类 > 文鱼型

  Fantail Goldfish > Original Type > Fantail Goldfish

- **品种详述** / Species Description

  蓝文鱼身体为梭形，具四开尾，眼睛、头部、鳞片等正常无变异。体色纯蓝无杂色，二龄后体色加深并具金属光泽，观赏价值较高。同时蓝文鱼在金鱼品种的杂交选育中，是比较好的一种载体。

  Blue Fantail Goldfish has fusiform spindle-shaped body, split caudal fin, and normal eye, head and scale. Pure blue. The color deepens and flashes like meta two years later, featuring high ornamental value. At the same time, Blue Fantail Goldfish is a good carrier for hybridization of goldfish.

雪青文鱼

Lilac Fantail Goldfish

花色类别 / Color •

雪青 / Lilac

品种类别 / Species •

文形 > 原型类 > 文鱼型

Fantail Goldfish > Original Type > Fantail Goldfish

品种详述 / Species Description •

　　雪青文鱼是近年来杂交选育的一个文鱼花色品种，体形比较修长，是体形比较标准的文鱼，配上雪青的典雅色调，给人以占朴简约之美。以雪青文鱼为亲本，在此基础之上，还可以选育出雪青文球等新品种。

Lilac Fantail Goldfish is a new variety developed in recent years. It has long body and standard figure. Cooperated with graceful violet color, the fish conveys the beauty of plain and simplicity. Lilac Fantail Goldfish with Pompons can be developed by using Lilac Fantail Goldfish as their parents.

红白文鱼 — Red White Fantail Goldfish

- **花色类别** / Color
  红白 / Red White

- **品种类别** / Species
  文形 > 原型类 > 文鱼型
  Fantail Goldfish > Original Type > Fantail Goldfish

- **品种详述** / Species Description

  红白文鱼是文鱼中红白花色的品种。图例中的红白文鱼全身洁白，唯独头顶有心形红斑，属于巧色，类似于锦鲤中的丹顶，比较受欢迎。

  Red White Fantail Goldfish is in red and white, as shown in the figure, it is with white body, and heart-shaped red spot on the head, which is ingenious and popular, very similar to Kio with red head pompon.

花色类别 / Color •

紫蓝 / Chocolate Blue

品种类别 / Species •

文形 > 原型类 > 文鱼型

Fantail Goldfish > Original Type > Fantail Goldfish

品种详述 / Species Description •

　　紫蓝文鱼同蓝文鱼系出同门，没有别的差异，只是在纯蓝体色上有铁锈色斑块。紫蓝文鱼有时会褪色变成蓝白、黑白和红白的个体，主要和温度、光照度及饲养水质软硬有关。

Chocolate Blue Fantail Goldfish shares origin with Blue Fantail Goldfish. Except rust spot on pure blue body, they have no differences. Chocolate Blue Fantail Goldfish may turn to blue and white, black and white, or red and white sometimes, depending on temperature, illuminance and feeding water quality.

紫蓝文鱼 —— Chocolate Blue Fantail Goldfish

蓝白文鱼 —— Blue White Fantail Goldfish

- **花色类别** / Color
  蓝白 / Blue White

- **品种类别** / Species
  文形 > 原型类 > 文鱼型
  Fantail Goldfish > Original Type > Fantail Goldfish

- **品种详述** / Species Description

  　　蓝白文鱼是蓝文鱼中逐渐褪色的个体，不具稳定性，但在褪色过程中，会让人领略到金鱼不断变化的一种美丽。

  Blue White Fantail Goldfish is Blue Fantail Goldfish individuals which lost color. It's not stable. However, during the process of change, we can perceive the beauty conveyed by the change of goldfish.

花色类别 / Color •
软鳞三色 / Calico

品种类别 / Species •
文形 > 原型类 > 文鱼型
Fantail Goldfish > Original Type > Fantail Goldfish

品种详述 / Species Description •

花文鱼
一
Calico Fantail
Goldfish

花文鱼是近年来杂交选育的一个文鱼花色品种，体形较其他花色文鱼要壮硕一些。图例为虎纹花色的品种，鱼鳍较宽，二龄后有较高的观赏价值。

Calico Fantail Goldfish is a recently developed Fantail Goldfish variety, featuring larger body than other types of Fantail Goldfish. The figure shows the variety with tiger strip and wide fins. It has high ornamental value after two years old.

文形
Fantail Goldfish

⌄

原型类
Original Type

⌄

和金型
Wakin

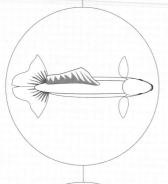

和金是日本金鱼的一个品种类型。近年来，随着中日金鱼品种的交流发展，目前国内也有饲养。和金身体外部形态和草金类似，也可以说就是日本的草金。和金的尾鳍将单尾（日本称之为鲋尾）、三开尾和四开尾都归纳在内。中国金鱼一般是将三开尾作为次品淘汰，而日本金鱼中的和金、土佐金却都是三开尾品种，可见两国金鱼文化虽然同源却又有不同的发展。和金体态强健，较易饲养，花色以红白居多，可以和草金一样作为庭院池塘中的观赏用鱼。

Wakin is a type of Japanese goldfish. Along with the increasing exchange between China and Japan in goldfish varieties, there are Wakin living in China at present. The appearance of Wakin is similar to Common Goldfish, in another word, Wakin is the Common Goldfish in Japan. According to the types of caudal fin of Wakin, it's further classified into single tail ("Funao"as per Japanese), triple tails and quadruple tails. In China, goldfish with three tails will be weeded out. But in Japan, Wakin and Tosakin are both of triple tails. Therefore, we can see that although China and Japan share the same origin of goldfish culture, they have developed towards different directions. Wakin is robust and easy to raise. Most Wakins are in red and white. Like Common Goldfish, it can be raised in the pool of courtyard as ornamental fish.

红白和金 | Red White Wakin

花色类别 / Color •

红白 / Red White

品种类别 / Species •

文形 > 原型类 > 和金型

Fantail Goldfish > Original Type > Wakin

品种详述 / Species Description •

　　图例所示为国内培育的和金。国内培育的和金与一般的草金比，体态更丰满强健，较易饲养，花色以红白居多，可以和草金一样作为庭院池塘中的观赏用鱼。和金野性较大，如果家庭中用较小容器饲养，易因突然受到惊吓或气压低缺氧时跃出饲养容器，因此，室内在用水族箱或水槽饲养时，最好加盖以防其跃出。

The figures show the Wakin cultivated in China. Compared with Common Goldfish, the Wakin cultivated in China is more plump and robust, and easy to raise. Most of the Wakins are in red and white, being able to serve as ornamental fish in the pool of courtyard. Wakin is of wild nature. When raising indoor, if the aquarium is small, it may jump out the aquarium due to being scared or lack of oxygen. Therefore, when raising indoor with aquarium or water channel, it's a must to cover the aquarium or water channel to prevent Wakin jumping out.

文形
Fantail Goldfish

∨

头型变异类
Head Variation

∨

鹅头型
Goosehead

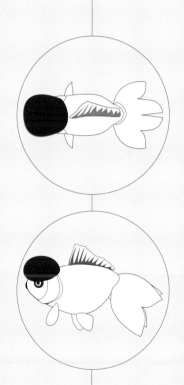

　　文形金鱼和蛋形金鱼中的头型变异类都可以划分为鹅头型、虎头型、狮头型和龙头型四大类型。其中，鹅头型的特征是头顶部具较发达顶茸，而头部其他区域相对比较正常、平滑，一般无吻凸。有的地区也会称之为高头或鹅冠。鹅头型名称源自鹅头红，因此，现将和鹅头红同样类型的头型变异类划归为鹅头型。

　　文形金鱼头型变异类中，鹅头型最具知名的代表品种莫过于鹤顶红，其他的如红色、红白花色和五花花色的金鱼目前饲养很少，甚至已经绝迹。鹅头型的顶茸要求质地细密规整，出檐而不遮目，面颊和鳃盖平滑干净。游动时稳健，静止时不倾头、不趴底。

Among Fantail Goldfish and Egg-Fish, head variation can be classified into Goosehead, Tigerhead, Lionhead and Dragonhead. The characteristics of Goosehead is developed pompon at the top of the head, while other parts of the head are relatively normal and smooth. There is no snout in general. In some area, such goldfish is called "tall head" or "goose crown". The name of Goosehead is originated from Goosehead Red Goldfish. Therefore, any head variation similar to Goosehead Red Goldfish is classified into Goosehead.

For head variations of Fantail Goldfish, the most famous is Goosehead. Other types such as red, red white, or five colors are rare and even extinct. The head pompon of Goosehead must be fine and regular, without covering the eyes, and with clean cheek and branchiostegite. It shall be vigorous when swimming, and not lean downwards or stay at bottom when not moving.

鹤顶红 | Red Cap Oranda (Goosehead Type)

- **花色类别** / Color

  红白 / Red White

- **品种类别** / Species

  文形 > 头型变异类 > 鹅头型

  Fantail Goldfish > Head Variation > Goosehead

- **品种详述** / Species Description

　　鹤顶红，又称一点红，是中国金鱼头型变异类中最具代表性的品种之一。其身银白似雪，鱼鳍轻如蝉翼，唯有头顶嵌入一颗宝石般的红冠，游动时典雅高贵，如同仙鹤飞舞，故称之为鹤顶红。鹤顶红纯化度较高，遗传稳定，其寓意吉祥，色彩搭配堪称完美，因此广受欢迎，中国各地鱼场都有饲育。优质的鹤顶红要求通体洁白，并有银质光泽，全身无杂色。鱼鳍舒展通透，无血丝。红冠色泽鲜艳，殷红如血，以头茸紧实，出檐而不遮眼，两颊素净，前不过唇，后不及背者为上品。鹤顶红鳞片质地较薄，如操作不慎非常容易脱落，欣赏价值会大打折扣。

Red Cap Oranda, also named "Yi Dian Hong", is one of the most representatives of head variations in China. It's silvery white like snow, with fins light as cicada's wings, and gem like red crown at the top of the head. When swimming, it's graceful, like flying crane, so it's also called "He Ding Hong". Red Cap Oranda is very pure and stable in heredity. As symbolizing good fortune, and perfect in color matching, it's widely accepted and cultivated in goldfish farms across China. Outstanding Red Cap Oranda is completely white with silvery gloss, and no mixed color at anywhere. The fins are stretching and clear, and no blood streak shall be seen. The crown is bright and red like blood. The head pompon is intensive and tight, stretching but without covering the eye, not reaching the lips in the front, and back in the rear. The cheeks shall be clean. The scale of Red Cap Oranda is very thin, which is easy to peel off if operating improperly, thus impair the ornamental value.

宽尾鹤顶红

Red Cap Oranda Broad Tail (Goosehead Type)

- **花色类别** / Color
  红白 / Red White

- **品种类别** / Species
  文形 > 头型变异类 > 鹅头型
  Fantail Goldfish > Head Variation > Goosehead

- **品种详述** / Species Description

　　宽尾鹤顶红又称裙尾鹤顶红，是近些年来，在传统鹤顶红的基础上，通过杂交选育的方式获得的一个品种。它的欣赏特点和传统鹤顶红类似，只是头茸变小，而尾叶变宽，且二龄以后尾叶下垂，如同白色长纱裙一般美丽，更加高贵典雅。

Red Cap Oranda Broad Tail (Goosehead Type) is a new variety obtained by hybridization based on traditional Red Cap Oranda. It shares same ornamental features as traditional Red Cap Oranda (Goosehead Type). However, it has smaller head pompon and wider tail leaf which will droop and look like long white veil when it's two years old, making it more graceful.

花色类别 / Color •
红白 / Red White

品种类别 / Species •
文形 ＞ 头型变异类 ＞ 鹅头型
Fantail Goldfish ＞ Head Variation ＞ Goosehead

品种详述 / Species Description •

红白高头其形态特征基本与鹤顶红类似，所不同的是颜色为红白花色，多居鹤顶红中挑选出来比较漂亮的红白高头。

Red White Oranda (Goosehead Type) is basically the same as Red Cap Oranda in shape. However, it has mixed red white color. They are selected Red Cap Oranda (Goosehead Type) with beautiful mixed color.

红白高头 | Red White Oranda (Goosehead Type)

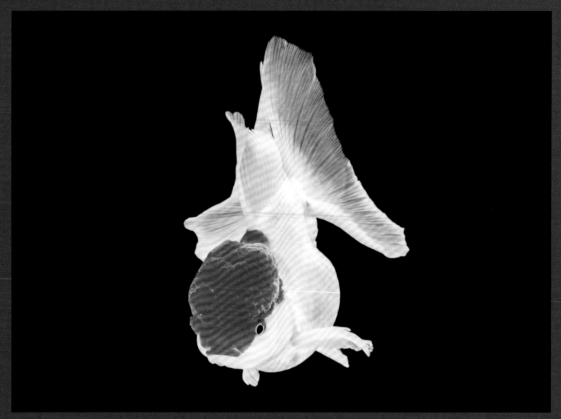

文形
Fantail Goldfish
∨
头型变异类
Head Variation
∨
虎头型
Tigerhead

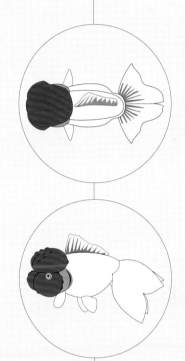

　　文形金鱼头型变异类中，虎头型所占比例较大，大多数传统上称为"帽子""高头"或"狮子头"的文形金鱼，都可以划入这个类别。虎头型的特征是头顶部具较发达的顶茸，鬓茸、目下和吻凸也较发达，而鳃盖区域相对比较正常、平滑。过去，南方和北方对金鱼中虎头和狮头的称呼一直较为混乱，主要是由于流传过程中的讹误，同时也没有人对头茸的分布做一个系统的归纳和梳理，只是简单地用有无背鳍来划分，这明显有误，因为不论是文形金鱼还是蛋形金鱼，其头型变异类中都会有虎头型和狮头型，而这两种头型是有明显差异的，与有无背鳍没有丝毫的关系。目前我们对虎头型和狮头型的定义，是以民国初的《金鱼丛谈》作为依据，它是目前可以看到最早有金鱼照片的图录，那时期，南北金鱼的命名还没有太大的差异与分歧。

In head variations of Fantail Goldfish, Tigerhead accounts for a large proportion. They are called "Hat", "Tall head" or "Lionhead" in ancient times. The characteristics of Tigerhead is dense head pompon, sideburns pompon, under the eye and snout. While the branchiostegites are relatively normal and smooth. In the past, the denominations of Tigerhead and Lionhead were confused, because there were mistakes during communications, and there was no one who had systematically concluded and sorted out the distribution of head pompon. The classification in the past was simply based on whether there was a dorsal fin. Obviously, it's not correct, as both the variations of Fantail Goldfish and Egg-Fish have Tigerhead and Lionhead. While such two head variations are different from each other in aspects not relating to the dorsal fin. At present, the definitions for Tigerhead and Lionhead are based on the *Goldfish Essay* written in the beginning of the Republic of China, which is the earliest publication which has pictures of goldfish available for now. At that time, there were no obvious differences and disputes between the denominations at the North and the South.

红狮头 | Red Oranda (Tigerhead Type)

- **花色类别** / Color
  红 / Red

- **品种类别** / Species
  文形 > 头型变异类 > 虎头型
  Fantail Goldfish > Head Variation > Tigerhead

- **品种详述** / Species Description

　　红狮头，又称大红狮，是传统金鱼中最具悠久历史的一个典型代表品种。其全身火红，身形雄健，饲养也较为容易，因此较为普遍，全国各地都有出产。但风格主要有南方和北方两个类型，图例所示为南方的红狮头，其特点为头茸适中，体高较宽，尾鳍宽大，比较适合侧视；其他各地均以北方风格居多，特点是头茸较南方的更发达夸张，但体高较南方窄，比较适合俯视。

Red Oranda (Tigerhead Type) also called Big Lionhead, is the representative of the traditional goldfish of the longest history. It is red all over the body, strong and easy to feed, so is very prevalent through out the country. Per the styles, there are mainly north type and south type. As is showing in the legend, it is a south Red Oranda (Tigerhead Type), with suitable head pompon, tall figure and wide caudal fin, so it is appropriate for side view; the rest belong to north style, which has an extremely developed head pompon, while it has narrow figure, so is better for top view.

白狮头 | White Oranda (Tigerhead Type)

• 花色类别 / Color
白 / White

• 品种类别 / Species
文形 > 头型变异类 > 虎头型
Fantail Goldfish > Head Variation > Tigerhead

• 品种详述 / Species Description

图例所示为白狮头。虽然习惯上称之为狮头，其实还是属于文形金鱼里的虎头型。白狮头一般为红白狮头中出的全白个体，这其中会有一些个体具红眼，习惯上称为朱砂眼或玉兔。而在一缸各色金鱼中，搭配一到两尾这样的朱砂眼金鱼，也是非常养眼的。白色金鱼一般鱼场不做过多保留，因此在市场上也只能偶然看到。

The figures show White Oranda (Tigerhead Type). Although it is called Lionhead customarily, it's in fact Tigerhead in Fantail Goldfish. White Oranda (Tigerhead Type) refers to the individuals of Red White Oranda (Tigerhead Type) which are completely white. Some of them with red eye are called "red eye" or "jade rabbit". One or two "red eye" in the aquarium will improve the overall effect. In general, goldfish farms do not keep much white ones, so there are few of white goldfish in the market.

蓝狮头 | Blue Oranda (Tigerhead Type)

- **花色类别** / Color
  蓝 / Blue

- **品种类别** / Species
  文形 > 头型变异类 > 虎头型
  Fantail Goldfish > Head Variation > Tigerhead

- **品种详述** / Species Description

　　蓝狮头饲养的地域甚多，而各地饲养的手法及金鱼的风格各有千秋。图例所示为福州培育的蓝狮头。福州蓝狮头头茸发达而不松散，三块分布，分界明显；身体雄健，尾鳍宽大舒展，有王者的气势，是非常优秀的金鱼品种。相对于福州蓝狮头，上海的蓝狮头的头茸略小，身体修长，但尾鳍较宽大。而徐州的蓝狮头虽然头茸较发达，但容易松散开花，身体较雄壮，尾鳍偏小。

Blue Oranda (Tigerhead Type) is generally raised in many regions with different ways of raising and styles. The figure shows the Blue Oranda (Tigerhead Type) cultivated in Fuzhou. Fuzhou Blue Oranda (Tigerhead Type) is outstanding goldfish type, featuring developed and intensive head pompon in three areas with distinct boundaries, robust, large and stretching caudal fin, and great momentum. Compared with Fuzhou Blue Oranda (Tigerhead Type), the head pompon of Shanghai Blue Oranda (Tigerhead Type) is relatively thinner and longer, with wider caudal fin. Though Xuzhou Blue Oranda (Tigerhead Type) has developed head pompn, it's easy to loose. It has robust body, but relatively small caudal fin.

雪青狮头 | Lilac Oranda (Tigerhead Type)

- **花色类别** / Color
  雪青 / Lilac

- **品种类别** / Species
  文形 > 头型变异类 > 虎头型
  Fantail Goldfish > Head Variation > Tigerhead

- **品种详述** / Species Description

　　图例所示为雪青狮头。雪青狮头曾经颇为流行，天津、江苏、上海、福建、广东等地均有饲养，但目前已不多见。雪青色金鱼色彩古朴而不绚丽，多属于小众花色品种，在一缸金鱼中多作为配色。图例中的雪青狮头，头茸发育适度，茸粒细腻饱满，不松散，身形矫健，尾鳍尾叶尖端虽有反折，但宽大舒展，各部比例匀称，是一尾比较优秀的金鱼。

The picture shows Lilac Oranda (Tigerhead Type), It was popular before and raised in Tianjin, Jiangsu, Shanghai, Fujian, Guangdong and so on. Yet, it is rare. The color of Lilac Oranda (Tigerhead Type) is plain and one from the small color range so it has a frequent role as a supporter. The Lilac Oranda (Tigerhead Type) in the picture has a moderate pilose hair with rich and tight particle. The robust figure, wide and stretched fish fin with a little opposite fold on the top are appropriated form of a stunning goldfish species.

红白狮头 | Red White Oranda (Tigerhead Type)

- **花色类别** ／ Color
  红白 ／ Red White

- **品种类别** ／ Species
  文形 > 头型变异类 > 虎头型
  Fantail Goldfish > Head Variation > Tigerhead

- **品种详述** ／ Species Description

  红白狮头，因南北地域习惯不同，也有叫做红白虎头的。传统出产于江苏的苏州、扬州和徐州等众多地区。其中，以顶茸紧实、体形雄健敦实、鱼鳍舒展宽大者为佳。红白狮头在花色上变化丰富，如类似图例中全身透红且鱼鳍不带白梢，而唯有顶茸雪白者称为玉印，古谱中也有雅称一捧雪或堆玉的，还有类似十二红的花色，都是红白狮头里的珍稀名贵品种，可遇而不可求。

Red White Oranda (Tigerhead Type) has another name of Red White Oranda (Tigerhead Type) according to different habbits from the Sorth and North. Its traditional born places are Suzhou, Yangzhou, Xuzhou, Jiangsu Province and etc. Among them, the fish with tight pilose head pompon, robust and tough figure and broad fish fin are the most valuable. Red White Oranda (Tigerhead Type) has splendid colors such as the overall red body with absent white fish fin while the specie with white pilose head pompon named jade seal. They also boast the name of a cluster of snow of a pile of jade. There is handful of rare species in Red White Oranda (Tigerhead Type) such as the specie whose color is the same as bombycilla japonica.

紫蓝狮头
Chocolate Blue Oranda
(Tigerhead Type)

- **花色类别** / Color
  紫蓝 / Chocolate Blue

- **品种类别** / Species
  文形 > 头型变异类 > 虎头型
  Fantail Goldfish > Head Variation > Tigerhead

- **品种详述** / Species Description

　　图例所示为紫蓝狮头，属于传统型狮头。传统狮头的头茸呈条带状分布，二龄以后容易松散或遮住眼睛。因此，在选育时尽量选择细茸粒或呈块状的。家庭饲养要控制好生长节奏，不要使头茸过分发育。紫蓝狮头有些会褪色成三色狮头，也是观赏性较强的一个花色品种。传统的紫蓝狮头体形适中，尾鳍不如福州狮头宽大，有待提高。

The pictures show Chocolate Blue Oranda (Tigerhead Type), a traditional variation of Oranda (Tigerhead Type). The head pompon of traditional Oranda is easy to be loosen or veil the eyes after the second instar. Therefore, goldfish with fine or massive pompon should be chosen as seed. Good growth pace should be kept in family rearing, the pompon should not be excessively developed.Some of the Chocolate Blue Oranda (Tigerhead Type) will fade into Tri-Colored Oranda (Tigerhead Type), another variation with high value of appreciation. Traditional Chocolate Blue Oranda (Tigerhead Type) are of medium build, their moors are not as broad as Oranda in Fuzhou, and there is still room for improvement.

三色狮头 | Red Black White Oranda (Tigerhead Type)

- **花色类别** / Color
  三色 / Red Black White

- **品种类别** / Species
  文形 > 头型变异类 > 虎头型
  Fantail Goldfish > Head Variation > Tigerhead

- **品种详述** / Species Description

  三色狮头以顶茸紧实、体形雄健敦实、鱼鳍舒展宽大者为佳。色彩搭配上总体要求颜色层次对比分明，白色银亮，黑色厚重，红色鲜亮，且呈块面状分布不散乱者为上。若出现红顶白肚黑背黑鳍者，则更是可遇不可求的逸品。

  Red Black White Oranda (Tigerhead Type) is favored for tight pilose head pompon, robust body, broad fish fins. The color is clearly comparable with different layers containing silver white, thick black and bright red and the color distribution in a chunk form is the most competitive one. The goldfish with red head pompon, white stomach, black back and black fin is the invaluable treasure.

云石狮头 —— Marble Oranda (Tigerhead Type)

- **花色类别** / Color
  软鳞黑白 / Marble

- **品种类别** / Species
  文形 > 头型变异类 > 虎头型
  Fantail Goldfish > Head Variation > Tigerhead

- **品种详述** / Species Description

　　云石狮头，也是近些年来选育的一个新的花色品种，出产于多地。和重彩狮头一样，它也是对传统狮头的花色和尾形加以改良并取得成功的一个新花色品种，俯视、侧视俱佳。在花色创新上它借鉴了云石神仙鱼的风格，软鳞的白色基底加上不同粗细黑色的条纹线条，如同传统中国水墨画一般，再辅以红色或橘色的顶茸，古谱中称为"旭日东升"或"霞光万道"，有一种柔和的新中国风之美。

Marble Oranda (Tigerhead Type) is also one of the newly breeding varieties which are produced in many places. Just like the Oranda (Tigerhead Type) (with Intense Colors), the Marble Oranda (Tigerhead Type) is also an excellent ornamental from overlook and side-look view after the improvement of its colors and caudal fin. For color innovation, it draws on its original style, adding different thickness of black stripe lines to its white Soft-Scale as the traditional Chinese ink paintings. In addition, the top color of the Marble Oranda (Tigerhead Type) is red or orange which means "red-sun rising" or "glittering rays of clouds" in the ancient spectrum endowed with the Chinese-style beauty of mellow softness.

水墨狮头 | Water Pattern Oranda (Tigerhead Type)

- **花色类别** / Color
  软鳞黑白 / Water Pattern

- **品种类别** / Species
  文形 > 头型变异类 > 虎头型
  Fantail Goldfish > Head Variation > Tigerhead

- **品种详述** / Species Description

　　图例所示为水墨狮头，是近几年来比较流行的软鳞花色品种。水墨狮头为软鳞黑白，色泽稳定，由软鳞五花狮头选育而出。与之类似的花色为云石，两者的区别是水墨黑多白少，而云石正好相反，是白多黑少，类似于白胜更纱和赤胜更纱的区别。图例2所示金鱼顶茸发育完美，身形健硕，是一尾形色俱佳的金鱼个体。

In the picture, it is Water Pattern Oranda (Tigerhead Type), is a color pattern relatively popular in recent years. Water Pattern Oranda (Tigerhead Type) has black and white Soft-Scale, steady color, their seeds are selected from Soft-Scale Calico Oranda (Tigerhead Type) (Bluish Base). Another similar design is marble, the difference between the two is that there are more black color than white color in Water Pattern Oranda (Tigerhead Type), while there are more white color than black color on Marble Goldfish (Tigerhead Type), just like the difference between Akagachisarasa and Shirogachisarasa. In figure 2, it has perfectly developed pompon, strong body and is a perfect individual with good shape and color.

虎纹狮头 — Tiger Banded Oranda (Tigerhead Type)

- **花色类别** / Color
  软鳞红黑 / Tiger Banded

- **品种类别** / Species
  文形 > 头型变异类 > 虎头型
  Fantail Goldfish > Head Variation > Tigerhead

- **品种详述** / Species Description

  虎纹狮头是由五花狮头中选育出的一个花色类型。五花色金鱼中的蓝色部分对水温水质较敏感，高温状态下蓝色易丢失，因此，南方的五花花色中较易出现虎纹花色。虎纹花色为显性遗传，基因稳定。传统金鱼中，虎纹花色一般不入品，但目前东南亚一带较为流行，国内也受此影响。

  Tiger Banded Oranda (Tigerhead Type) is a variation selected from Calico Oranda (Tigerhead Type) (Bluish Base). The blue color in the calico goldfish is sensitive to the temperature and quality of water. The blue color is easy to get lost under high temperature. Therefore, calico goldfish tend to grow in the south, furthermore, tiger stripe has dominant and steady genetic model. Goldfish with Tigerhead don't belong to traditional goldfish, but are prevalent in Southeast Asia, China is also affected.

撒锦狮头 | Oranda (Tigerhead Type) (with Bluish Matt and Black Dots)

- **花色类别** / Color
  软鳞五花 / Bluish Matt and Black Dots

- **品种类别** / Species
  文形 > 头型变异类 > 虎头型
  Fantail Goldfish > Head Variation > Tigerhead

- **品种详述** / Species Description

　　撒锦狮头的花色又称芝麻花，是软鳞五花中分离出的一个花色品种，其特点是软鳞鱼，体色为白色或淡蓝色，部分有硬质反光鳞。全身分布碎点状黑花，如同撒上的黑芝麻粒。图例所示为福州产撒锦狮头，体格雄壮，花色素雅，头茸发育适度，较适合大型水族箱侧视观赏。

Oranda (Tigernhead Type) (with Bluish Matt and Black Dots), also called sesame flower, is a variation detached from Soft-Scale Calico Oranda (Tigerhead Type). The body is white or bluish, many with hard reflecting scales, dotted with black flowers that look like scattering black sesame dots. The ones shown here are Oranda (Tigerhead Type) (with Bluish Matt and Black Dots) produced in Fuzhou, with strong body, light color, moderate head pompon, very appropriate to keep in large-scale side-view aquarium.

麒麟狮头
— Kirin Oranda
(Tigerhead Type)

花色类别 / Color •

软鳞五花 / Kirin

品种类别 / Species •

文形 > 头型变异类 > 虎头型
Fantail Goldfish > Head Variation > Tigerhead

品种详述 / Species Description •

　　图例所示为麒麟狮头。它是金鱼业者近几年选育而出的、比较受市场欢迎的一个新花色品种。麒麟狮头是软鳞五花色中出现的全身青灰色或青黑色的个体，腹部呈现白色，其鳞片一般基部颜色较深，边缘处较浅，排列整齐，十分富有层次感。身体纯青灰色称为墨麒麟，而如图例身上有大块色斑的称为花麒麟。麒麟花色如果配以红顶者，会显得更加高贵。麒麟是传说中的瑞兽，因此现在比较受欢迎。

The picture shows the Kirin Oranda (Tigerhead Type) which is a rather popular variety among fish market as a newly breeding fish.The Kirin Oranda (Tigerhead Type) is one of the calico varieties of Soft-Scale whose body is totally blue gray or black with white abdomen.The base color of its scale is generally deeper, while the edge color is rather lighter, neat and rich. The fish with pure gray body is called Dark Kirin while the fish with large spots is called Flower Kirin which is shown in the picture. The Flower Kirin will appear more noble with red top because Kirin is a legendary auspicious which is very popular nowadays.

重彩花狮 —— Oranda (Tigerhead Type) (with Intense Colors)

- **花色类别** / Color

  软鳞五花 / Intense Colors

- **品种类别** / Species

  文形 > 头型变异类 > 虎头型

  Fantail Goldfish > Head Variation > Tigerhead

- **品种详述** / Species Description

重彩花狮，是金鱼业者近些年来选育的一个新的花色品种，也广受市场好评。其顶茸紧实、体形雄健敦实。尤其是其尾鳍，在吸取泰狮优点的基础上，对传统狮头的尾形加以改良并取得成功，成为俯视侧视俱佳的一个品种。花色创新上采用重彩，也是对传统色彩审美的一种突破，软鳞的基底加上厚重的色调，如同传统重彩水墨画一般，对欣赏者有极强的视觉冲击力。

Oranda (Tigerhead Type) (with Intense Colors) is one of the newly breeding fish varieties and also gains good reputation among the fish market. The head pompon of Oranda (Tigerhead Type) (with Intense Colors) is tight and tip while its body shape is stocky, especially for its caudal fin, has been improved on the basis of the advantages of the traditional Thailand Oranda and creates an perfect ornamental from overlook and side-look view. Intense colors are employed in the innovation of color of the Oranda (Tigerhead Type) (with Intense Colors), which is a breakthrough of the traditional color aesthetics as well. Besides, the Soft-Scale of the Oranda (Tigerhead Type) (with Intense Colors) is added with strong colors like traditional Chinese ink paintings that has a strong visual impact on viewers.

五花狮头

Calico Oranda (Tigerhead Type) (Bluish Base)

- **花色类别** / Color

  软鳞五花 / Calico (Bluish Base)

- **品种类别** / Species

  文形 > 头型变异类 > 虎头型

  Fantail Goldfish > Head Variation > Tigerhead

- **品种详述** / Species Description

　　五花狮头是著名的传统金鱼品种，全国各地都有饲养，现在比较著名的出产地有天津、唐山、南京、苏州、徐州、福州等。前面5个产地基本风格相似，都属于传统五花狮头。而现在福州的五花狮头却是自成一格。图例1所示为传统五花狮头，头茸非常发达，又称菊花顶。图例2、3为福州培育的五花狮头，其吸收了泰狮的流行风格，体形巨大，头茸紧实，尾鳍舒展飘逸，如持彩练当空舞。后期如能将素蓝色提纯出来，将会成为非常优秀的中国金鱼代表品种。

Calico Oranda (Tigerhead Type) (Bluish Base) is one of the famous traditional goldfish breeding varieties all over the country. The famous breeding places include Tianjin, Tangshan Nanjing, Suzhou, Xuzhou, Fuzhou etc. The basic type of the Calico Oranda (Tigerhead Type) (Bluish Base) appears the same in the fore five places which belongs to the traditional types. However, Calico Oranda (Tigerhead Type) (Bluish Base) in Fuzhou is very unique. As is shown in the picture 1, the traditional Calico Oranda (Tigerhead Type) (Bluish Base) has a developed head pompon which is also called the "chrysanthemum top". The picture 2 and picture 3 show the Calico Oranda (Tigerhead Type) (Bluish Base) breeding in Fuzhou which has a huge body, a compact head pompon and soft caudal fin like a flying ribbon in the sky after absorbing the popular style of Thailand Oranda. Later, it will become a very good representative of the Chinese goldfish varieties if people can purify the pigment blue.

文形
Fantail Goldfish

∨

头型变异类
Head Variation

∨

狮头型
Lionhead

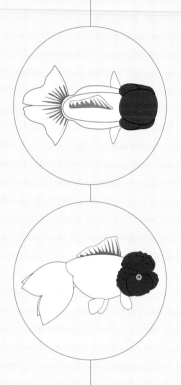

　　文形金鱼和蛋形金鱼的头型变异类都可以划分为鹅头型、虎头型、狮头型和龙头型四大类型。文形金鱼中，狮头型是近些年来，福州金鱼业者培育发展出来的一个新的大系列，并有广阔的市场空间。狮头型主要是国外一些金鱼品种引进后杂交选育，也有的是原有的传统品种发生变异后加以提纯稳定，有的则是利用寿星或福寿中出现的文形金鱼个体加以固化选育。因此，与文形金鱼中的虎头型变异不同，狮头型变异在外部形态上有较大差异。在文形金鱼中狮头型头茸分布和蛋形金鱼中狮头型的头茸分布基本一样，所以使用相同的头型定义，即整个头部都有头茸分布，十分发达，如同雄狮威武的狮鬃一般，故名之曰狮头。文形金鱼中的狮头型，大多体格健壮，易于饲养，大多数可以发育成体形巨大的金鱼。文形金鱼狮头型的培育，是南派金鱼业者对中国金鱼的巨大贡献。

Among Fantail Goldfish and Egg-Fish, head variation can be classified into Goosehead, Tigerhead, Lionhead and Dragonhead. Lionhead in Fantail Goldfish is a major innovation of goldfish industry in Fuzhou, and of great potential market. Lionhead is developed by several means, such as hybridization with foreign goldfish varieties, purifying and stabilizing variation of original species, or selection and cultivation of Fantail Goldfish individuals in Lionhead (Southern) or Ranchu. Therefore, unlike Tigerhead in Fantail Goldfish, Lionhead variations are varying in appearance. The head pompon of Lionhead in Fantail Goldfish is similar to that of Lionhead in Egg-Fish. Therefore, the definition of head for the two is the same, i.e., splendorous pompon across the head like the mane of martial male lion. This is also the origin of their name. Most Lionhead in Fantail Goldfish are robust and easy to raise, and have big body after grown-up. The cultivation of Lionhead in Fantail Goldfish is a great contribution of southern goldfish industry for Chinese goldfish.

花色类别 / Color •

黑 / Black

品种类别 / Species •

文形 > 头型变异类 > 狮头型
Fantail Goldfish > Head Variation > Lionhead

品种详述 / Species Description •

黑狮头
Black Oranda
(Lionhead Type)

　　黑狮头是文形金鱼狮头型中，比较受欢迎的一个品系。传统风水学认为，黑色金鱼可以阻挡煞气，因此，饲养黑狮头有趋吉避凶的寓意。图例所示黑狮头头茸发育适中，整体感强；尾鳍上翘舒展，尾鳍末端出现类似日本土佐金尾鳍上的朝颜，十分美丽。整体比例协调、稳健，总体风格上吸收借鉴了泰狮的身形特点，但比泰狮多出几分端庄娴静。

Black Oranda (Lionhead Type) is a popular type of Lionhead in Fantail Goldfish. As per traditional Fengshui theory, black goldfish is capable to stop bad luck. Therefore, raising Black Oranda (Lionhead Type) is to pursue good fortune and avoid disaster. The figure shows a beautiful Black Oranda (Lionhead Type) which has temperate development, fine overall look, and upwards stretching caudal fin with Pharbitis at the tip (similar to Asagao goldfish). Its overall posture is balanced and steady, as absorbed the merits of posture of Thailand Oranda, however more demure.

蓝狮头

Blue Oranda
(Lionhead Type)

- **花色类别** / Color
  蓝 / Blue

- **品种类别** / Species
  文形 > 头型变异类 > 狮头型
  Fantail Goldfish > Head Variation > Lionhead

- **品种详述** / Species Description

　　图例所示为蓝狮头，小尾型。福州当地也称呼这种体形浑圆、小尾巴的狮头型为元宝狮。元宝狮的体形在福州金鱼当中属于小身形的一类，突出特点就是圆身短尾，似乎可以托在手掌中把玩。相对于其他蛋形金鱼狮头型的霸气，元宝狮体现的是小拙的可爱。元宝狮体形不大，要求养得精致，对头型、鳞片、色彩具有较高的要求。

The figures show the Blue Oranda (Lionhead Type) with small tail. In Fuzhou, such round and small tailed Lionhead is also called Yuanbao Oranda (Lionhead Type). Among Fuzhou goldfish, Blue Oranda (Lionhead Type) is relatively small. Its prominent characteristics are round and short tail, enabling playing in palm. Compared with other martial Lionhead in Egg-Fish, Yuanbao Oranda (Lionhead Type) shows clumsy and cute. The size of Yuanbao Oranda (Lionhead Type) is small, thus requires special care and high requirements on head, scale and color.

紫扯旗兰寿 | Chocolate Ranchu with Dorsal Fin

- **花色类别** / Color
  紫 / Chocolate

- **品种类别** / Species
  文形 > 头型变异类 > 狮头型
  Fantail Goldfish > Head Variation > Lionhead

- **品种详述** / Species Description

　　图例所示为紫扯旗兰寿。紫扯旗兰寿是紫兰寿中出现的带背鳍的返祖个体，其在外形上同蛋形金鱼的紫兰寿没有太大区别，只是多了一个背鳍。紫扯旗兰寿的饲养也同普通国寿一样，可以饲养在较深的水族缸里，适合侧视。其身体雄壮、体质强健，较易饲养，但受众不多。

What the picture shows is Chocolate Ranchu with Dorsal Fin. With a dorsal fin on back ,it is a variant of Chocolate Ranchu and it dose not possess lots of difference from Egg-Fish, compared with Chocolate Ranchu. The feeding method is just like common that of National Ranchu. In addition, It could feed in aquarium with deep water and seeing from the side could provide a better view. With strong physical quality, it is very easy to feed but cannot enjoy a great popularity among costumers.

花色类别 / Color •

雪青 / Lilac

品种类别 / Species •

文形 > 头型变异类 > 狮头型

Fantail Goldfish > Head Variation > Lionhead

品种详述 / Species Description •

雪青扯旗兰寿
Lilac Ranchu
with Dorsal Fin

　　图例所示为雪青扯旗兰寿，是雪青色国寿中出现带背鳍的返祖个体。其头茸发育适度、规整。体形健硕，全身淡雪青色且无杂斑。传统金鱼业者一般对扯旗兰寿接受度不高，而其主要是满足一些金鱼爱好者求新求怪的猎奇心理。因此，一般鱼场只是在正常国寿中发现这些返祖个体，保留一些，而不做专门培育。

What the picture shows is Lilac Ranchu with Dorsal Fin, it is a variant of Lilac Ranchu. It has fare head pomon, strong body, and is lilac without any spots. But it is not well received among traditional goldfish raisers, only for fanciers of curiosity. Therefore, in general fish ground, it could merely be discovered among common National Oranda, and kept for small range without special cultivation.

红白狮头 — Red White Oranda (Lionhead Type)

- **花色类别** / Color
  红白 / Red White

- **品种类别** / Species
  文形 > 头型变异类 > 狮头型
  Fantail Goldfish > Head Variation > Lionhead

- **品种详述** / Species Description

图例所示为红白狮头，和其他所有金鱼一样，红白花色大多是该品系中最受欢迎的一个花色类别。红白狮头中，花色变化很多，亦很漂亮。如图例1、6、7中的通背红，日本称之为鲣节。图例2中玉面，图例3玉顶，又称一捧雪，日本称为大窗。此外还有十二红、六鳞红、卷腹红、鹿子红等等，都是红白狮头中特别优秀的花色。

The figures show Red White Oranda (Lionhead Type). Like all other goldfish, red and white is the most popular classification in this variation. The color variation of Red White Oranda (Lionhead Type) is rich and beautiful. For example, the red back shown in figure 1, 6 and 7 is called Katsuobushi in Japan, the white face in figure 2 and white top in figure 3 is called Jade Head, or Oomado as per Japanese. Moreover, there are many other excellent color variations of Red White Oranda (Lionhead Type), such as Twelve Red, Rokurin, Red Belly, as well as Scarlet Scale, and so on.

红黑狮头 Red Black Oranda (Lionhead Type)

- **花色类别** / Color
  红黑 / Red Black

- **品种类别** / Species
  文形 > 头型变异类 > 狮头型
  Fantail Goldfish > Head Variation > Lionhead

- **品种详述** / Species Description

　　图例所示为红黑狮头，是文形金鱼中的黑狮头，逐渐转向红色体色的过渡阶段的个体。其体色是过渡色，因此不稳定，大多数黑色金鱼都会经历这样一个变色过程变为红色金鱼。其外形特征与蛋形金鱼狮头一样，也分大尾类型和圆身短尾的元宝狮类型，体质强，较易饲养。

What the picture reveals is Red Black Oranda (Lionhead Type). It is one type of Fantail Goldfish with Black Oranda (Lionhead Type), which is transforming into red color. As its color is in the process of transforming, the color is not very stable. The majority of black golden fish will go through this stage to become red golden fish. Its shape is just like other Egg-Fish and is also classified into the type with big fishtail and the type with round body and little fishtail. It is strong physically and easy to feed.

紫红狮头

Chocolate Red Oranda
(Lionhead Type)

- **花色类别** / Color
  紫红 / Chocolate Red

- **品种类别** / Species
  文形 > 头型变异类 > 狮头型
  Fantail Goldfish > Head Variation > Lionhead

- **品种详述** / Species Description

　　图例所示为紫红狮头。江浙一带饲养的文形金鱼虎头型中也有类似的花色，称之为朱顶紫罗袍，其曾经在香港被拍卖出天价，并引起轰动。此类花色一般为紫色向红色转色的过渡色，尚不稳定，需要继续做提纯稳固。目前家庭饲养可以在转色时降低水温，在饲料中添加螺旋藻等手段延缓其变色过程。

The pictures show Chocolate Red Oranda (Lionhead Type). Some types of Fantail Goldfish with Tigerhead that fed in Jiangsu and Hangzhou province also possess this color, which is called Red Head with Chocolate Luopao and it was once bid with high price in Hong Kong, which was pretty a big news at that time. Generally speaking, the color is not stable as it is transforming from purple to red, and the further enhance and refinement for the stability of color is also necessary. Currently, there are some method for the stability of color, such as turning down the water temperature and adding spirulina in fish food.

黑白狮头 —— Black White Oranda (Lionhead Type)

- **花色类别** / Color
  黑白 / Black White

- **品种类别** / Species
  文形 > 头型变异类 > 狮头型
  Fantail Goldfish > Head Variation > Lionhead

- **品种详述** / Species Description

　　图例所示为黑白狮头，是金鱼业者近些年来培育的金鱼新品种，也有人称之为短尾狮头。它经由蓝元宝狮选育而出，转为黑白花色。其圆身短尾，头茸特别发达。一般蓝色转色出的黑白色或喜鹊花色稳定性较差，不能持久。家庭饲养可以在转色时降低水温，在饲料中添加螺旋藻等手段延缓其变色过程。图例所示的黑白狮头头茸发育未控制好，二龄后会松散。

As shown in the picture, it is Black White Oranda (Lionhead Type), newly cultivated by goldfish fanciers in recent years, local people sometimes call it Lionhead Short Tail. It is cultivate from Blue Yuanbao Oranda when its color changes into black white. It is with round body and short fishtail as well as developed head pompon. The black white color or magpie color out of the blue color is not stable and lasting. During the color transition, in order to keep the color duration, it is better to keep a low water temperature and add spirulina into the fish food. The one in the picture has a poor head pompon, will get loosen after 2 years old.

蓝白狮头 ——Blue White Oranda (Lionhead Type)

- **花色类别** / Color
  蓝白 / Blue White

- **品种类别** / Species
  文形 > 头型变异类 > 狮头型
  Fantail Goldfish > Head Variation > Lionhead

- **品种详述** / Species Description

　　图例所示为蓝白狮头，其体态匀称、丰满，尾鳍宽大且舒展；头茸发育适度，不松散，有较好的发展空间；颜色为蓝白，是蓝色向白色转色的一个过渡花色，日本称为羽衣。蓝白色若能稳固会是一个优秀的花色类型，但目前尚未能做到。狮头型的蓝白狮头是近几年培育出的一个品种，较受市场欢迎。

As is showing here is Blue White Oranda (Lionhead Type). It has round and plump body, with wide and stretching caudal fin; the head pompon is moderate developed, and has room to become better; the blue white color is the transition color from blue to white. Japanese people name it Hagoromo. The blue white color is regarded excellent if it's stable, which is not so far until now. Blue White Oranda (Lionhead Type) is newly cultivated in recently years, and is quite popular in market now.

紫蓝狮头 | Chocolate Blue Oranda (Lionhead Type)

- **花色类别** / Color
  紫蓝 / Chocolate Blue

- **品种类别** / Species
  文形 > 头型变异类 > 狮头型
  Fantail Goldfish > Head Variation > Lionhead

- **品种详述** / Species Description

　　图例所示为紫蓝狮头。紫蓝狮头全身青蓝色，和蓝狮头相同，但其鱼鳍多为紫色，身上也带有紫色斑块。狮头型的紫蓝狮头也是近些年来才出现的品种，其风格也是借鉴了泰狮，走大体形、宽尾鳍的路线。这种紫蓝狮头培育时间不长，还没有完全达到成熟的标准，仍要不断改良。

The picture is Chocolate Blue Oranda (Lionhead Type). Its entire body is blue green, just like Blue Oranda with chocolate fins. Some may also carry a few purple spots on body. It is also a new type that is created in recent years. Combined some feature of Thailand Oranda, it is bigger with wide fins. The cultivation time of Chocolate Blue Oranda (Lionhead Type) is not long and it hasn't reached mature standard, so, further improvement is still necessary.

花色类别 / Color •

红白 / Red White

品种类别 / Species •

文形 > 头型变异类 > 狮头型
Fantail Goldfish > Head Variation > Lionhead

品种详述 / Species Description •

　　图例所示为红白扯旗兰寿。红白扯旗兰寿是福寿中出现的带背鳍的返祖个体，在过去金鱼养殖场多将其作为次品淘汰，而现在，为了迎合市场上一些金鱼爱好者求新求怪的猎奇心理，将这些带背鳍的国寿保留下来饲养。红白扯旗兰寿在外形上同蛋形金鱼的红白兰寿没有太大区别，只是多了一个背鳍，一般蛋形金鱼中出现的全背鳍会比正常文形金鱼的背鳍偏小，因此称之为扯旗。

The figures show Red White Ranchu with Dorsal Fin, it is a reversion of Ranchu which has dorsal fin. In the past, such goldfish would be weeded out in goldfish farm. However, for meeting the novel interests of certain goldfish amateurs, such Ranchu with dorsal fin is kept. Red White Ranchu with Dorsal Fin is similar to Red White Ranchu, but with a dorsal fin. The full featured dorsal fin of Egg-Fish is generally smaller than normal　Fantail Goldfish, so it's called "Cheqi".

红白扯旗兰寿
Red White Ranchu with Dorsal Fin

三色狮头 | Red Black White Oranda (Lionhead Type)

- **花色类别** / Color
  三色 / Red Black White

- **品种类别** / Species
  文形 > 头型变异类 > 狮头型
  Fantail Goldfish > Head Variation > Lionhead

- **品种详述** / Species Description

　　图例所示三色狮头，是近些年来金鱼业者选育的一个新品种类型，与传统三色狮头金鱼相比区别在于，其头型属于狮头型。根据尾鳍又可分为小尾和大尾两种类型，小尾的显得憨拙，大尾的风格雄健。其共同特点是体高较宽，比较适合水族箱内侧视观赏。

The pictures show Red Black White Oranda (Lionhead Type), it is also one new type cultivated in recent years. Compared with traditional Red Black White Oranda (Tigerhead Type), it belongs to Lionhead is pretty special. Based on caudal fin, it could be classified into small tail and big tail. The fish with small tail is sort of clumsy and that with big tail is very lusty. Their common feature is their height is pretty high, so they are suitable for watching in aquarium.

樱花狮头 | Sakura Oranda (Lionhead Type) (Red White Matt)

- **花色类别** / Color

  软鳞红白 / Red White Matt

- **品种类别** / Species

  文形 > 头型变异类 > 狮头型

  Fantail Goldfish > Head Variation > Lionhead

- **品种详述** / Species Description

　　樱花狮头是文形金鱼狮头型中，软鳞红白色的一个花色品种。樱花花色的鉴赏，一般以白多红少为佳品；其红色部分要求最好是殷红色，而呈现绛红色的更为难得。白色硬质反光鳞片，宜少不宜多。图例2所示的樱花狮头顶茸非常发达，短尾类型，花色上红色部分显橙红色，且面积过大，反光鳞片偏多，与图例3、4相比，花色要略逊一筹。

Sakura Oranda (Lionhead Type) (Red White Matt) is Red White Soft-Scale among Fantail Goldfish. According to the appreciation of Sakura, the fish with more white than red are the great one; the fish with deep red would be the better one than that with common red color, and the fish with magenta would be the best. The white Hard-Scale could reflect light, so the fish with less Hard-Scale can be the better ones. In figure 2, it is Sakura Oranda (Lionhead Type) (Red White Matt) with well-developed head pompon and short fishtail. Its red color is too much and it tends to be orange and there are too much reflective scales. Therefore, its color is not as good as the ones in picture 3 and 4.

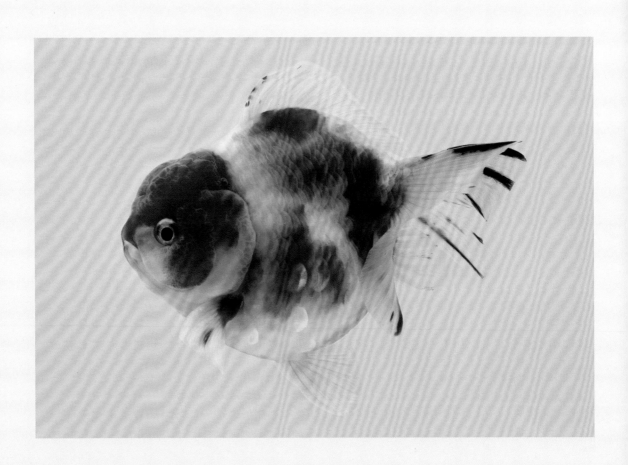

撒锦狮头——Oranda (Lionhead Type) (with Bluish Matt and Black Dots)

- **花色类别** / Color
  软鳞五花 / Bluish Matt and Black Dots

- **品种类别** / Species
  文形 > 头型变异类 > 狮头型
  Fantail Goldfish > Head Variation > Lionhead

- **品种详述** / Species Description

　　图例所示为撒锦狮头。撒锦花色过去俗称芝麻花，是软鳞五花中分离出的一个花色类型，其特点是软鳞鱼，体色为白色或淡蓝色，部分有硬质反光鳞，全身分布碎点状黑花，如同撒上的黑芝麻粒。图例所示为福州产撒锦狮头，体格雄壮，花色素雅，头茸发育适度，根据尾鳍大小，又可分为大尾和小尾两个不同品系。撒锦最适合与麒麟配合一起饲养，二者颜色对比强烈，且都较适合大型水族箱侧视观赏。

The picture shows Oranda (Lionhead Type) (with Bluish Matt and Black Dots). This color was called sesame color, Soft-Scale Calico (Bluish Base). Its feature is Soft-Scale, with white or light blue colors and Hard-Scale which could reflect light. What's more, the entire body is covered black flower, just like black sesame. The picture is the Oranda (Lionhead Type) (with Bluish Matt and Black Dots) fed in Fuzhou City. With proper head pompon and elegant caudal fin, it is strong and decorated with clean flower. There are big-tail type and small-tail type according to the sie of caudal fin. This type could be fed together with Kirin Oranda, as their colors could create contrast view and both of them could be enjoy in big aquarium.

花狮头 | Calico Oranda (Lionhead Type)

- **花色类别** / Color
  软鳞五花 / Calico

- **品种类别** / Species
  文形 > 头型变异类 > 狮头型
  Fantail Goldfish > Head Variation > Lionhead

- **品种详述** / Species Description

　　图例所示花狮头，是近些年来金鱼业者选育的一个新品种类型。其头茸适中、体形雄健敦实；尾鳍在吸取泰狮优点的基础上，对传统狮头的尾形加以改良并取得成功，成为俯视侧视俱佳的一个品种。花色上虽不及重彩来得绚丽夺目，倒也朴实厚重。

The pictures show Calico Oranda (Lionhead Type), it is also a new type with proper head pompon and strong body. Absorbing the feature of Thailand Oranda, its caudal fin successfully improved, which enable it to be a wonderful type for watching from above. Its color, although not as splendid as Lionhead (Intense Color), is very pure and dignified.

花色类别 / Color •

软鳞五花 / Kirin

品种类别 / Species •

文形 > 头型变异类 > 狮头型
Fantail Goldfish > Head Variation > Lionhead

品种详述 / Species Description •

麒麟狮头 | Kirin Oranda (Lionhead Type)

　　图例所示为麒麟狮头，是金鱼业者近几年选育出的比较受市场欢迎的一个花色品种。麒麟狮头是从软鳞五花品种中选育而出的，其鳞片一般基部颜色较深，边缘处较浅，排列整齐，十分富有层次感。身体纯青灰色称为墨麒麟，而如图例身上出现色斑或碎花的，称为花麒麟。麒麟花色如果配以红顶者，会显得更加高贵。麒麟是传说中的瑞兽，因此麒麟花色的金鱼，现在比较受欢迎。

The pictures show Kirin Oranda (Lionhead Type), which is a very popular variety cultivated by goldfish raisers. Developing from Soft-Scale Calico, the color of its scale generally is sort of dark at the bottom, then gets lighter at edge organized in good order. The color spots showing in the picture is called Calico Kirin. The Calico Kirin if growing with red head pompon, would become more glorious. Kirin is an auspicious monster in the legend. Therefore, the fishes with this color are very popular among people presently.

重彩狮头 —— Oranda (Lionhead Type) (with Intense Colors)

- **花色类别** / Color
  软鳞五花 / Intense Colors

- **品种类别** / Species
  文形 > 头型变异类 > 狮头型
  Fantail Goldfish > Head Variation > Lionhead

- **品种详述** / Species Description

　　图例所示重彩狮头，是近些年来金鱼业者选育出的一个新品种类型。其头茸适中、体高较宽、尾鳍多为小尾类型，显得憨态可掬、朴拙可爱。其花色为软鳞重彩类型，如果有适宜的灯光效果，浓重的色彩搭配上反光鳞片，给人以炫目瑰丽的感觉，是比较适合水族箱内侧视观赏的一个品种。

The pictures show the Oranda (Lionhead Type) (with Intense Colors), a new breeding cultivated in recent years. It is generally with moderate head pompon, wide size and small caudal fin, very clumsy and cute. Their color is intense and scale is soft. With proper lights, intense color and scale can reflect light, creating an amazing appreciation effect. It is suitable for watching from two sides of aquarium.

五花狮头 ｜ Calico Oranda (Lionhead Type) (Bluish Base)

- **花色类别** / Color
  软鳞五花 / Calico (Bluish Base)

- **品种类别** / Species
  文形 > 头型变异类 > 狮头型
  Fantail Goldfish > Head Variation > Lionhead

- **品种详述** / Species Description

  　　图例所示为五花狮头，是近些年来金鱼业者在传统五花狮头的基础上，改良选育出的一个新品种类型。其尾鳍较传统五花狮头显得宽大。花色选取蓝色作为底色，配以红色、黄色、黑色、白色的组合。

  The pictures show Calico Oranda (Lionhead Type) (Bluish Base), selected and bred from traditional Calico Oranda (Lionhead). Its caudal fin is pretty big than traditional variety and its color is mainly blue, with white, yellow, black and white.

文形
Fantail Goldfish

ⓥ

头型变异类
Head Variation

ⓥ

龙头型
Dragonhead

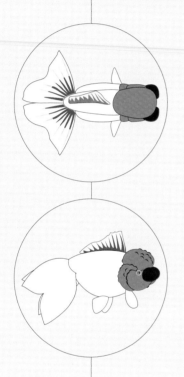

　　文形金鱼中鹅头型、虎头型和狮头型都是传统划分类型，而龙头型是一个全新的概念。之所以提出龙头型，是因为近几年来，日本金鱼以及泰国金鱼被引入国内饲养，而它们的头型标准，用传统的划分方法去定义不太准确，故而引入龙头型这个概念。龙头型这个名词，原来是日寿头型划分中的一个类型，在文形金鱼中，我们把头型类似日寿的这种龙吻特别发达、头型俯视呈长方形的变异类型称之为龙头型。龙头型顶茸为方形，发育度适中，尤其龙吻外凸，十分显眼。鬓茸及鳃茸较薄，这一点介于虎头型与鹅头型之间，但鹅头型吻部不凸出，而虎头型的吻部也不及龙头型发达，俯视整体呈四方形或圆形。龙头型对头茸的发育要求较高，要求规整紧实，如雕刻出来的感觉，绝对不能松散、杂乱无章。

Goosehead, Tigerhead and Lionhead belong to traditional classification of Fantail Goldfish. However, Dragonhead is totally new concept. The reason why we put up with this concept is that, in recent years, with the importing of Japanese and Thailand goldfish, they cannot be classified into traditional types, thus Dragonhead was created. Dragonhead is one type of Japanese Ranchu. When the Fantail Goldfish seems like Japanese Ranchu with big snout and square head with top view, it is called Dragonhead, whose pompon is moderate in square shape and the snout is extremely outstanding. The sideburns pompon and gill pompon are not very bushy, just between Tigerhead and Goosehead. But the snout of Goosehead does not bulge out and the snout of Tigerhead is not as developed as Dragonhead, when viewed from top, it is square or round. The Dragonhead is highly demanded for the cultivation of pompon, which has to be neat and firm, just like sculpture without any mess and clutter.

花色类别 / Color •

软鳞五花 / Calico, Red White Matt

品种类别 / Species •

文形 > 头型变异类 > 龙头型

Fantail Goldfish > Head Variation > Dragonhead

品种详述 / Species Description •

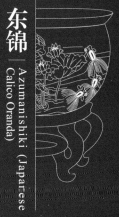

东锦 | Azumanishiki (Japanese Calico Oranda)

　　东锦是日本金鱼，近些年来刚刚被引入国内饲养。有资料记载，东锦是通过荷兰狮头杂交选育的。其体色为软鳞五花或软鳞樱花；头型为近似日寿的龙头型，龙吻非常发达，尾鳍宽大，尾肩平展，非常富有张力；体形中等，体质较弱，对水质等要求较高，较一般国内金鱼难饲养。东锦目前在国内还属于小众品种，饲养的也多为日系金鱼爱好者，多作俯视欣赏。

Azumanishiki (Japanese Calico Oranda) is a type of Japanese goldfish and was introduced in China in recent years. According to relevant data records, it is cultivated through the hybridization of Dutch Oranda and its body carries the color of Soft-Scale Calico (Bluish Base) or Sakura with Soft-Scale. Its head is very similar with the Dragonhead Japanese Ranchu and its snout is very developed. What's more, its caudal fin is wide and its middle tail is flat and streching. With medium size, it is weak and highly demand for water quality, so it is difficult to feed in China. This type of goldfish is merely a niche and only Japanese goldfish fans are fond of feeding them. It could be appreciated from above.

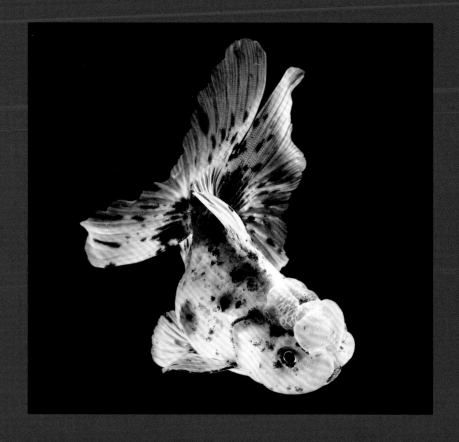

软鳞五花 / Calico, Red White Matt

花色类别 / Color •

红白 / Red White

品种类别 / Species •

文形 ＞ 头型变异类 ＞ 龙头型

Fantail Goldfish ＞ Head Variation ＞ Dragonhead

品种详述 / Species Description •

日
狮

Oranda Shishigashira
(Japanese Oranda)

　　图例所示为日狮，也称红白日狮，和东锦一样，也是近些年来刚刚引入国内市场的，时间略晚于东锦。日狮外形特征上与东锦非常接近，其头型也属于典型的龙头型。日狮体色为红色或红白色，硬质鳞。饲养界对日狮尾鳍的要求也比较严格，尾肩平展，且要有力度，尾鳍宽大且富张力。日狮身形偏长，头身比可达1：3～1：2.5。日狮性格较胆怯，不易与人亲近。

The pictures show Oranda Shishigashira (Japanese Oranda), it is also called Red White Japanese Oranda, imported to China after AzumanishikiIt in recent years. It is similar to Azumanishiki in figure and possesses typical Dragonhead. Its color is red or white with Hard-Scale. It has high standard for the shape of caudal fin, flat, wide with strong tension and middle tail, flat and stretching. In addition, its body is comparatively long and the proportion between head and body could reach 1:2.5-3. However, it is kind of shy and not willing to get close to people.

文形
Fantail Goldfish

∨

眼型变异类
Eye Variation

∨

龙睛型
Dragon Eyes

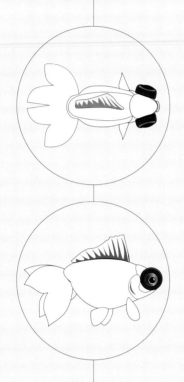

　　龙睛是中国传统金鱼的代表，因其双目外凸，和神话传说中龙的眼睛非常相似，故而得名。华夏民族一直以龙为自己民族的图腾，因此，以龙命名，可见对它的推崇。龙睛出现得非常早，明朝中后期便有记述，很多民间的艺术作品中都有龙睛的形象。1960年6月1日，我国邮电部发行的特38《金鱼》主题邮票，一套12枚，其中4枚便是各种花色的龙睛。龙睛曾经被单独列出来划为一个大类，与草、文、蛋并列，称为中国金鱼四分法，由此可见，龙睛在中国金鱼中的地位之高。龙睛在国内各地均有饲养，比较有特点的是徐州的大眼龙睛和西安的蚕豆眼龙睛。龙睛的花色也有很多，常见的有红龙睛、墨龙睛、蓝龙睛、紫龙睛、花龙睛、红白龙睛、五花龙睛等等。由龙睛发展出来的复合变异类金鱼也很多，如龙睛蝶尾、龙睛球、龙睛高头、龙睛高头球等等。龙睛虽然有着辉煌的历史，但在现今却逐渐凋零，其逐渐被龙睛蝶尾所取代。

Dragon Eye is a representative of traditional Chinese goldfish. The eyes protrude and resemble the eyes of dragons in myths and legends, hence the name comes out of that. The Chinese has always used dragon as a totem of the nation, so naming after dragon can show people's worship for dragons. Dragon Eyes was born very early and recorded in the middle and later Ming Dynasty. In many folk works of art, there were images about them. On June 1, 1960, the Ministry of Posts and Telecommunications of China issued a set of 12 special "38" Goldfish themed stamps, 4 of which were an assortment of Dragon Eyes. It used to be listed as a single species, to distinguish from Grass (Crucian), Fantail Goldfish and Egg-Fish, collectively known as a quartering of Chinese goldfish. From this, it can be seen that Dragon Eyes has a high position among Chinese goldfish. It is bred in various areas in China. The most characteristic ones are Xuzhou Big Dragon Eyes and Xi'an Bean Dragon Eyes. It has many colors. The common ones are Red Dragon Eyes, Black Dragon Eyes, Blue Dragon Eyes, Chocolate Dragon Eyes, Calico Dragon Eyes, Red White Dragon Eyes, Calico Dragon Eyes (Bluish Base), etc. Since there are many composite variants of Dragon Eyes, such as Butterfly Moor, Dragon Eyes with Pompon, Oranda with Dragon Eyes, and Dragon Eyes Pompon Oranda, etc. Although it has a glorious history, it has gradually faded away today and been replaced by Butterfly Moor.

红龙晴
—Red Telescope

花色类别／Color •

红／Red

品种类别／Species •

文形 ＞ 眼型变异类 ＞ 龙晴型
Fantail Goldfish ＞ Eye Variation ＞ Dragon Eyes

品种详述／Species Description •

　　图例所示为红龙晴。红龙晴大约出现在明朝中晚期，是历史悠久的古老金鱼品种之一。1960年我国邮电部发行的特38《金鱼》邮票，其中第12枚就是红龙晴。红龙晴玩赏以中长体形配合长尾为最佳。红龙晴体质强健，易于饲养，是金鱼爱好者饲养初期练手的最佳选择。

The figures show Red Telescope. Red Telescope emerged in the middle and late Ming Dynasty and is one of the most time-honored goldfish varieties. In the special "38" Goldfish themed stamps issued by the Ministry of Posts and Telecommunications of China in 1960, the twelfth stamp was exactly Red Telescope. The best ornamental Red Telescope is medium or long-bodied and long-tailed. Red Telescope has a strong physique and is easy to feed. It is the best choice for beginners in the initial stage.

花色类别 / Color •

黑 / Black

品种类别 / Species •

文形 > 眼型变异类 > 龙睛型
Fantail Goldfish > Eye Variation > Dragon Eyes

品种详述 / Species Description •

黑龙睛
Black Telescope

　　黑龙睛是龙睛中的代表，又称墨龙，全身漆黑如墨，眼球膨大凸出于眼眶之外，一般为长尾，尾叶较窄。黑龙睛培育时间较长，因此颜色比较稳定。过去对黑龙睛的玩赏非常讲究，除了全身黑如墨染，连鱼嘴张开后，内里都要是黑色的。黑龙睛的主要玩赏点在眼球，大而匀称是基本要求。黑龙睛中的优秀代表是西安出产的黑蚕豆眼，目前已不多见。

Black Telescope is a representative of Dragon Eyes and also known as Black Moor. It is pitch-black from head to tail. The eyeballs protrude out of the orbit. Generally, it has a long tail and a narrow tail leaf. The cultivation time of Black Telescope is long, thus its color is stable. Previously, people were very fastidious about ornamental Black Telescope. It must be black from head to tail, and black inside when the mouth is opened. The main highlight of Black Telescope is its eyeballs. The basic requirements are big and symmetric. An outstanding representative of Black Telescope is Xi'an Black Bean Eyes, which is quite rare now.

红黑龙睛

— Red Black Telescope

- **花色类别** / Color
  红黑 / Red Black

- **品种类别** / Species
  文形 > 眼型变异类 > 龙睛型
  Fantail Goldfish > Eye Variation > Dragon Eyes

- **品种详述** / Species Description

　　红黑龙睛是黑龙睛逐步向红龙睛转色时的过渡体色个体，其花色也称包金色。1960年我国邮电部发行的特38《金鱼》邮票，其中第2枚黑背龙睛便是红黑龙睛。红黑龙睛在外部形态上与黑龙睛完全一样。龙睛的体形一般偏长，多以长鳍为佳。除了蚕豆眼眼型外，红黑龙睛还有苹果眼、算珠眼、牛角眼等眼型。

Red Black Telescope is a transitional species from Black Telescope to Red Telescope. The color is also called over gilt. In the special "38" Goldfish themed stamps issued by the Ministry of Posts and Telecommunications of China in 1960, the second Black-backed Dragon Eye was Red Black Telescope. It has exactly the same shape as Black Telescope. Dragon Eyes often has a long body, preferably with long fin. Apart from the Broad Bean Eyes, Apple Eyes, Abacus Bead Eyes and Bull Eyes, etc. are also seen in Red Black Telescope.

文形
Fantail Goldfish

⋁

眼型变异类
Eye Variation

⋁

蛤蟆眼型
Frog-Head

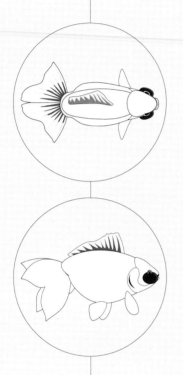

蛤蟆眼初名猴面，文形金鱼和蛋形金鱼中都有，因其眼部下方有硬质泡囊，极似蛙类的声囊，故而又名蛙头。蛤蟆眼是一个古老的金鱼品种，也有把它划分在头型变异类中的。但目前金鱼鉴赏中，单独对蛤蟆眼做玩赏的几乎没有，大多数情况是进行水泡的杂交育种时，作为一种中间的过渡体来使用，因此，将其归入眼型变异的范畴中较为妥当。蛤蟆眼的泡囊较小，内有淋巴液；头部较宽，有不太发达的头茸分布；体质强健，较易饲养。蛤蟆眼属于小众品种类型，数量稀少，可见到的花色也不是很多，主要是红白和五花两种。

The Frog-Head used to be called monkey face in China. It exists in both Fantail Goldfish and Egg-Fish. Since it has a hard vesicle in the sub orbital region and a vocal sac similar to frogs, thus it is also called Frog-Head. It is an ancient goldfish variety and some people categorize it into the head variation class. But in current goldfish appreciation, few people appreciate the Frog-Head alone. In most cases, it is hybridized with Bubble-Eye and used as a transitional type. Thus, it would be more proper to categorize it into the eye variation. The Frog-Head has a small vesicle with lymph in it. The head pomon is wide and covered with a less developed cap. It has a strong physique and is easy to feed. The Frog-Head is a minority species and very scarce. They don't have many colors, but mainly red-and-white and streaky.

花色类别 / Color •

红白 / Red White

品种类别 / Species •

文形 > 眼型变异类 > 蛤蟆眼型
Fantail Goldfish > Eye Variation > Frog-Head

品种详述 / Species Description •

　　图例所示为红白蛤蟆眼。其头宽嘴阔，两眼有朱砂色虹彩；眼球下方与眼眶之间白膜膨胀鼓囊，泡囊内有淋巴液；泡囊轻微破损后，有一定的自修复能力，但如果完全撕裂后，则无法修复。红白蛤蟆眼较易饲养，以鱼盆、木海等俯视观赏为主，侧视观赏性不佳。

The figures show a Red White Frog-Head with Dorsal Fin. It has a wide head and broad snout. Both eyes have vermilion iris. There is a hard vesicle between sub eyeball region and orbit, with lymph in it. If the vesicle is slightly damaged, the Frog-Head can repair by itself. But if the vesicle is completely torn apart, it can't be repaired. Red White Frog-Head with Dorsal Fin is easy to feed and is mainly appreciated in a fish bowl or wood tub from top view. Side-looking doesn't have a good appreciation effect.

红白蛤蟆眼 — Red White Frog-Head with Dorsal Fin

文形
Fantail Goldfish

∨

眼型变异类
Eye Variation

∨

水泡型
Bubble-Eye

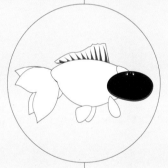

　　文形金鱼中的水泡眼是较老的一个金鱼品种，目前已非常少见。文形金鱼中的水泡眼虽然也有称之为扯旗水泡的，但它并非是蛋形金鱼水泡眼中带背鳍的返祖个体稳定固化而来的，而是一直都有的一个老品种。以前北京金鱼中的墨色水泡、蓝色水泡多为文形品种。目前国内，文形金鱼中的水泡眼最具代表性的产地为安徽合肥。合肥逍遥津公园在新中国成立初期就有繁育，迄今已有几十年的饲养历史，而且花色齐全，以五花色最具观赏性，其泡大且柔软，鱼鳍长。湖南也出产文形水泡，最具特色的是三色文形水泡，但数量也较稀少。文形水泡体质较强，较易饲养。

Bubble-Eye is an old variety of Fantail Goldfish, but is very rare now. Although being called also Red Bubble-Eye with Dorsal Fin, it is not stably inherited from an atavistic individual with dorsal fin among Bubble-Eye in Egg-Fish, but an old variety that always exists. Black Bubble and Blue Bubble in the past of Beijing goldfish was Fantail Goldfish. At present, the most typical origin of Bubble-Eye of Fantail Goldfish is Hefei, Anhui. Hefei Xiaoyaojin Park has bred it since the beginning of the founding of new China. So far, it has decades of breeding history. There are a full range of colors and calico (bluish base) is the most ornamental. It has large soft bubbles and long fins. Bubble-Eye in Fantail Goldfish are produced in Hunan, too. The most distinctive is tri-colored Bubble-Eye in Fantail Goldfish, which is very scarce. Bubble-Eye in Fantail Goldfish has a strong physique and are easy to feed.

红扯旗水泡
Red Bubble-Eye
with Dorsal Fin

花色类别 / Color •
红 / Red

品种类别 / Species •
文形 > 眼型变异类 > 水泡型
Fantail Goldfish > Eye Variation > Bubble-Eye

品种详述 / Species Description •

红扯旗水泡是文形金鱼水泡型中较常见的花色，但饲养地区较少，以安徽为主，上海也有少量出产。安徽的扯旗水泡身体稍纤细，水泡大而柔软，鱼鳍长而飘逸，遗传也相对稳定。图例所示为上海产的红扯旗水泡，为杂交选育的品种，体态粗壮，泡体相对较小，尾型为小宽尾，显得憨拙可爱。

Red Bubble-Eye with Dorsal Fin is a common color in Fantail Bubble-Eye. But it is only reared in a few areas, principally Anhui and produced in Shanghai in small quantities. Anhui Red Bubble-Eye with Dorsal Fin has a slim body, large and soft bubbles and long and ethereal fins. The inheritance is stable, too. The figure shows a Shanghai Red Bubble-Eye with Dorsal Fin. It is a hybridized breed, with a sturdy body, small bubbles and a small and wide tail and looks silly but lovely.

紫扯旗水泡
Chocolate Bubble-Eye with Dorsal Fin

花色类别 / Color •

紫 / Chocolate

品种类别 / Species •

文形 > 眼型变异类 > 水泡型
Fantail Goldfish > Eye Variation > Bubble-Eye

品种详述 / Species Description •

　　紫水泡较少见，而紫扯旗水泡则更加珍稀。图例所示紫扯旗水泡泡囊较小，内有淋巴液，头宽嘴阔，头顶部有顶茸，二三龄后可发育为鹅头水泡；背鳍完整，尾鳍宽大飘逸；身体壮硕，体质强，较易饲养。紫扯旗水泡和大多数水泡一样，较不耐缺氧，故需低密度饲养。多补充高蛋白饵料可促进头茸及水泡发育，二三龄后达到最佳观赏效果。

Chocolate Bubble-Eye with Dorsal Fin is rare, but Chocolate Bubble-Eye with Dorsal Fin among Fantail Goldfish is even rarer. The figures show Chocolate Bubble-Eye with Dorsal Fin. It has a small vesicle with lymph in it. Also it has a wide head, broad mouth and pompon on its head. At the age of 2 or 3, the hood can be developed into raspberry-like hood. The dorsal fin is complete. The caudal fin is wide and ethereal. It has a strong physique and is easy to feed. Like a vast majority of bubbles, Chocolate Bubble-Eye with Dorsal Fin is less tolerant to anoxia, so it must be raised at a low density. Replenishing protein-rich bait can promote the development of pompon and bubble and achieve the best ornamental effect at the age of 2 or 3.

红黑扯旗水泡

Red Black Bubble-Eye with Dorsal Fin

- **花色类别** / Color

  红黑 / Red Black

- **品种类别** / Species

  文形 > 眼型变异类 > 水泡型

  Fantail Goldfish > Eye Variation > Bubble-Eye

- **品种详述** / Species Description

  　　图例所示为红黑扯旗水泡，属于杂交选育出的品种。其色不稳定，会逐步过渡为全红色，通过饲养手法的调整，可以延缓这个褪色过程。水泡二龄以后，多数会起顶茸，而此鱼系当岁鱼，顶茸发育较好，如有较稳定遗传，可将其划为文形金鱼复合变异类中的鹅头水泡类型。

  The figures show a Red Black Bubble-Eye with Dorsal Fin, bred by hybridization. The colors are not stable and will gradually change into all red. The discoloration process can last by adjusting the breeding technique. At the age of 2, most Bubble-Eye will develop a pompon. This fish is exactly at the age of 2, so head pompon is well-developed. If inherited stably, it can be categorized as raspberry-like Goosehead Bubble-Eye in composite variation of Fantail Goldfish.

文形
Fantail Goldfish

∨

鼻膜变异类
Nasal Variation

∨

文球型
Pompons

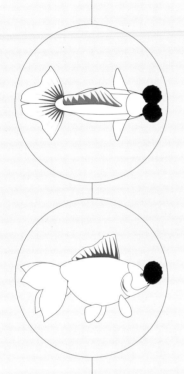

文形金鱼中，有鼻瓣膜特别发达、呈皱褶绒球状变异的，称为文球。文球为传统金鱼，因两绒球聚于头顶，如同插着两朵花，所以又雅称簪花。1960年，我国邮电部发行的特38《金鱼》邮票中第1枚，便是文球。文球过去一直较受欢迎，而现在由于饲养的数量减少，逐渐有些没落。文球有双球和四球之分。双球较多，球大而紧实；四球球体异常发达，游动时动感极强，但球体易松散，有时会被金鱼吸入口中，将球柄越拉越长，影响观赏效果。因此，文球应适时对绒球部分加以修剪。文形金鱼绒球的球体为皱褶瓣膜，有自修复能力，但如果球柄也损伤，则恢复较困难，且恢复时间需要较长。文球的花色类型较多，有红文球、白文球、紫文球、青文球、蓝文球、雪青文球、紫身红球、红白文球、樱花文球、五花文球等等。由文球还可以变异出文形金鱼高头球型等复合变异类品种。文球体质好，较易饲养，全国各地金鱼场均有繁育。

Among Fantail Goldfish, a variation has an especially developed nasal valve, which presents a wrinkled pompon variation, known as Fantail Goldfish with Pompons. Fantail Goldfish with Pompons is a traditional goldfish. Since two pompons are gathered on the top of head, just like two flowers, Fantail Goldfish with Pompons has an elegant name-hair flower. In the special "38" Goldfish themed stamps issued by the Ministry of Posts and Telecommunications of China in 1960, the first stamp was exactly a Fantail Goldfish with Pompons. In the past, Fantail Goldfish with Pompons were always popular, but now very few are bred and they gradually declined. Fantail Goldfish with Pompons can be divided into double- pompon and quadruple- pompon. Most are double- pompon. The pompons are large and compact. Quadruple-pompon is extremely developed and very dynamic when the fish swims. But the pompons are loose and sometimes sucked into the goldfish's mouth. The pompon stipe is elongated and affects the ornamental effect. Therefore, the pompons must be trimmed timely. The pompons of Fantail Goldfish have wrinkles and can repair themselves. But once the stipe is damaged, it will be difficult to repair and the repair takes a lot of time. Fantail Goldfish with Pompons has a lot of colors, such as Red Fantail Goldfish with Pompons, White Fantail Goldfish with Pompons, Chocolate Fantail Goldfish with Pompons,Green Fantail Goldfish with Pompons, Blue Fantail Goldfish with Pompons, Lilac Fantail Goldfish with Pompons, Chocolate Bodied Fantail Goldfish with Red Pompons, Red White Fantail Goldfish with Pompons, Sakura Fantail Goldfish with Pompons and Calico Fantail Goldfish with Pompons (Bluish Base), etc. Fantail Pompon can evolve into Fantail High Pompon and other composite mutants. Fantail Goldfish is bred in goldfish farms throughout China. It has a strong physique and is easy to feed.

花色类别 / Color •

红 / Red

品种类别 / Species •

文形 > 鼻膜变异类 > 文球型
Fantail Goldfish > Nasal Variation > Pompons

品种详述 / Species Description •

红文球
—— Red Fantail Goldfish with Pompons

　　图例所示为红文球。红文球多为大尾，偶尔有小尾品种，但多以四球大尾者为贵。红文球球体紧实，身形壮硕。其性格较活泼，易与人亲近，互动性好；在群体中游动速度快，摄食能力强，不宜与珍珠、水泡等混合饲养。南方饲养的形体高而宽（如图例1所示）。而北方饲养的体形略偏长。红文球头部尖，顶部多伴生薄薄一层顶茸。

The figures show Red Fantail Goldfish with Pompons. It is mostly large-tailed, and occasionally small-tailed. Quadruple-pompon Goldfish is more precious with large tail. Red Fantail Goldfish with Pompons has compact pompons and a strong physique. It has a lively and easy-going character and interacts with people well. It swims fast in the group and has strong ingestion ability. It had better not be fed with pearl or Bubble-Eye, etc. Red Fantail Goldfish with Pompons raised in the south has a tall and wide body, as shown in Figure 1, while Red Fantail Goldfish with Pompons raised in the north has a long body. Red Fantail Goldfish with Pompons has a pointed head, on which there is a thin layer of head pompon.

青文球 — Green Fantail Goldfish with Pompons

- **花色类别** / Color
  青 / Green

- **品种类别** / Species
  文形 > 鼻膜变异类 > 文球型
  Fantail Goldfish > Nasal Variation > Pompons

- **品种详述** / Species Description

　　青文球体色也有称为铁色，易与紫文球混淆，主要是幼苗期青文球为青黑色，而紫文球幼苗期为半透明黄色。青文球有转色个体，多转为黑色，少部分转为红色，而紫文球多会转为红色。图例1所示为双红色绒球，还有双白色绒球，图例2所示为一红一白的鸳鸯球。青文球和红文球一样，易于饲养。

The body color of Green Fantail with Pompons is also called iron color, and it may be confused with chocolate color. The main differences between Green Fantail with Pompon and Chocolate Fantail Goldfish with Pompons are: the fries of the former one are blue black while the fries of the latter one are translucent yellow. Green Fantail Goldfish with Pompons has color-changed individuals, most of which change to black and few to red, but most of Chocolate Fantail with Pompon change to red. In Suzhou, more Green Fantail Goldfish with Pompons are bred, including double-red floccule, as shown in figure 1, and double-white floccule, as shown in figure 2, and also red-and-white floccule. Green Fantail Goldfish with Pompons is as easy to breed as Red Fantail Goldfish with Pompons.

紫文球

Chocolate Fantail Goldfish with Pompons

- **花色类别** / Color
  紫 / Chocolate

- **品种类别** / Species
  文形 > 鼻膜变异类 > 文球型
  Fantail Goldfish > Nasal Variation > Pompons

- **品种详述** / Species Description

　　紫文球是目前文球中饲养最为广泛的一个花色品种。由于各地水质等因素的差异，紫色的表现力各不相同，以硬质微碱性水饲养出的最好；偏软的弱酸性水质饲养的紫文球颜色较暗，容易与青文球混淆。图例4所示为紫文球中最漂亮的紫身四红球。紫身红球全身赤铜色，唯独绒球现出红色，由于金鱼业者不断地提纯，紫身红球的颜色相较其他紫红色鱼更趋稳定，因此大受市场欢迎。紫身红球中有的偶尔会褪色变成紫红白三色文球，非常珍稀。

Chocolate Fantail Goldfish with Pompons is a variety that is bred most widely among Fantail Goldfish with Pompons at present. Due to the differences in water quality in different places, the expressive forces of chocolate are also different; and those bred in hard slightly alkaline water are the best. The color of those bred in softer slightly alkaline water is are darker, so it may be confused with Green Fantail Goldfish with Pompons. Among Chocolate variety, the most beautiful one is Chocolate Body with Four Red Pompons (as shown in figure 4). Chocolate Body with Red Pompons is red copper in the whole body, except for its floccule are is red. Because of constant refinement of the goldfish farmers, the color of this breed is more stable than that of other chocolate red breeds, so it is very popular in the market. Among Chocolate Body with Red Pompons, Chocolate Red White Fantail Goldfish with Pompons accidentally appears because of color fading, which is very precious and rare.

红白文球 — Red White Fantail Goldfish with Pompons

- **花色类别** / Color
  红白 / Red White

- **品种类别** / Species
  文形 ＞ 鼻膜变异类 ＞ 文球型
  Fantail Goldfish ＞ Nasal Variation ＞ Pompons

- **品种详述** / Species Description

　　图例所示为红白文球。红白文球较为珍贵，饲养数量稀少，尤以全身洁白如玉，唯独两个绒球鲜红和眼睛带虹彩的朱砂球、类似六鳞红的巧色红白文球特别珍稀罕见，可遇不可求。红白文球以双红球者为贵，而球体半红半白的花球，亦十分俏皮。红白文球体形修长，再配以如纱裙般的长阔尾，游动时尤显曼妙高贵。

The figures show Red White Fantail with Pompons, which is precious with small quantity. Especially, Cinnabar Pompons which is as white as polished jade in the whole body with two bright red floccules and irisated eyes, and the type in ingenious color, similar to Rokurin are very precious and rare. Among this variety, the ones with two red floccules are more precious, those with half-red half-white floccules are also very cute. The Red White Fantail Goldfish with Pompons is slender and looks beautiful and noble with dress-like long and broad tail when swimming.

红黑文球

Red Black Fantail Goldfish with Pompons

- **花色类别** / Color

  红黑 / Red Black

- **品种类别** / Species

  文形 > 鼻膜变异类 > 文球型

  Fantail Goldfish > Nasal Variation > Pompons

- **品种详述** / Species Description

  　　图例所示为红黑文球。红黑文球是黑文球向红文球转色时期的过渡期品种，因此不稳定。在黑色向红色转色时，偶尔会出现全身乌黑，而唯独两个绒球为红色的个体，非常罕见。如能定向培育将其颜色稳固下来，将是非常优秀的花色品种。红黑文球体质强健，身长尾长，易于饲养，适当降低水温可以延长褪色时间。

  The figures show Red Black Fantail Goldfish with Pompons, which is a transitional variety from black ones to red ones, so it is instable. When the color transits to red from black, the individuals that are jet-black in the whole body with two red floccules will appear occasionally and they are very rare. If the color can be stabilized through directive breeding, they will become a very excellent variety of Fantail Pompon Goldfish. Red Black Fantail Goldfish with Pompons is very strong, with long body and long tail, and is easy to breed. Appropriate low water temperature can extend the color fading time.

五花文球
Calico Fantail Goldfish with Pompons (Bluish Base)

- **花色类别** / Color
  软鳞五花 / Calico (Bluish Base)

- **品种类别** / Species
  文形 > 鼻膜变异类 > 文球型
  Fantail Goldfish > Nasal Variation > Pompons

- **品种详述** / Species Description

　　图例所示为五花文球。五花文球饲养数量不多，比较珍贵，过去扬州饲养较多，目前安徽培育较多。五花文球以红顶朱球、蓝背白腹，身上带黑碎点的芝麻花配色最为漂亮，但十分罕见。五花文球身体修长，鳍长如水袖舞动，头顶部有时伴生有顶茸，观赏时侧视俯视俱佳。五花文球身体强健，较易饲养，是文球中一个优秀的花色品种。

The figures show Calico Fantail Goldfish with Pompons (Bluish Base), which is precious and quantity is small, it was mainly found in Yangzhou, now is mainly raised in Anhui. Among this variety, the sesame blossom color combination of red-top floccules, blue back and white belly with black breaking point is the most beautiful, but very rare. It is slender, with long fins waving like dancing sleeves and pompon grows on the head sometimes. It is better to appreciate from sides and top. Calico Fantail Goldfish with Pompons (Bluish Base) is strong and easy to breed, and it is an excellent variety among Fantail Goldfish with Pompons.

文形
Fantail Goldfish

∨

背峰变异类
Hump Variation

∨

琉金型
Ryukin

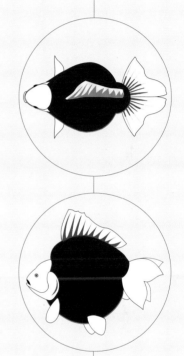

琉金是目前国内国际金鱼市场最受欢迎的金鱼品种之一。现代琉金的起源，通常认为是由中国的文鱼经古琉球国传到日本，由日本人逐步培育出的一个品种，因为是从琉球引入日本的金鱼品种，故称之曰琉金。琉金最大的特点是尖头高身，体形圆如盘状，很像神仙鱼，特别适合侧视。琉金从外形上分为短尾琉金、长尾琉金和宽尾琉金三大类别。其花色也是各种金鱼之中最多的一个品种。琉金体质强健，易于饲养，尤其适合大水面深水养殖，因此，国内许多地区采用大面积坑塘饲养。目前广东和福建出产的品质较高，全国其他各地也有大规模饲养，但因为气候环境、生长周期等诸多因素，其总体品质不如上述这两个地区。琉金与留金谐音，也有聚财的寓意，因此，不少人将它作为风水鱼饲养，加上它抗病性强，相对其他金鱼易于饲养，所以多年来一直长盛不衰。

Ryukin is one of the most popular goldfish varieties in the domestic goldfish market at present. Usually, it is believed that modern Ryukin is a variety gradually developed by Japanese after Chinese Fantail Goldfish was introduced to Japan via ancient Ryukyu Kingdom, and because it is goldfish variety introduced from Ryukyu, it is called Ryukin. The most important feature of Ryukin is sharp head, long body and disc-like body shape, and it looks like angel fish, and is suitable for appreciating from sides. In terms of appearance, Ryukin can be divided into short-tail Ryukin, long-tail Ryukin and wide-tail Ryukin. Its colors are also the richest among various kinds of goldfish. Ryukin is strong and is easy to breed; especially, it is suitable for deep-water breeding in large water, so in China, large-area pond breeding is adopted in many areas. At present, the major regions that produce Ryukin in China is Guangdong and Fuzhou. And it is also bred in other areas in China in large scale, but because of many factors, such as climatic environment and growth cycle, their overall quality is inferior to those in the above two areas. Since the Chinese characters of Ryukin and Liu Jin (Liu Jin, meaning retaining gold) are homophonic, so it has the connotation of gathering wealth. Therefore, many people breed it as a Fengshui Goldfish. Besides, it has a strong disease resistance and is easily to breed than other varieties of goldfish, so it has always been everlasting for many years.

红短尾琉金 — Red Ryukin Short Tail

- **花色类别** / Color

  红 / Red

- **品种类别** / Species

  文形 > 背峰变异类 > 琉金型

  Fantail Goldfish > Hump Variation > Ryukin

- **品种详述** / Species Description

  　　红短尾琉金分软鳞和硬鳞两种，硬鳞较常见，而软鳞多为樱花琉金中出现的全红个体。软鳞红短尾琉金通体红润，硬质反光鳞点缀其间；身圆头尖，背峰高耸，尾鳍短小而有力。短尾琉金家庭饲养要注意控制密度不宜过高，深水饲养为佳，北方冬季最好采取加温措施，否则低温容易造成其失鳔症，影响观赏。饲育时可多食用增体饲料。

  Red Ryukin Short Tail can be divided into two groups, i.e. Soft-Scale and Hard-Scale. The latter is more common. Soft-Scale Red Ryukin Short Tail is mainly Red Sakura Ryukin Short Tail (Red White Matt). The Soft-Scale Red Ryukin Short Tail is usually red, while the Hard-Scale one is dotted with reflective scales; it has a round body, sharp head, high back hump and short but powerful tail fin. When domestically breeding Ryukin Short Tail, the feeder should pay attention to control the density, which should not be too high; it is better to breed it in deep water. In winter, heating measures should be taken. Otherwise, low temperature can easily lead to loss of swim bladder, affecting the appreciation. Therefore, more body-reinforcing feed can be used.

雪青短尾琉金

Lilac Ryukin Short Tail

- **花色类别** / Color

  雪青 / Lilac

- **品种类别** / Species

  文形 > 背峰变异类 > 琉金型

  Fantail Goldfish > Hump Variation > Ryukin

- **品种详述** / Species Description

  图例所示为雪青短尾琉金。雪青短尾琉金是从蓝短尾琉金中选育的，为近年来新培育的一个花色品种。和大多数雪青色金鱼一样，多为小众花色品种，一般受到高级玩家的追捧。雪青短尾琉金和一般短尾琉金在外形上没有太大差异，其饲养也相对容易。雪青色褪色比例不是很高，多在一缸鱼中起到配色的作用。

  The figures show Lilac Ryukin Short Tail, which is selected from Blue Ryukin Short Tail, a new breeding in recent years. Like most of lilac goldfish, it is appreciated by a minor group and is generally pursuit by senior fanciers. In appearance, there is not great differences from common Ryukin Short Tail, and it is easier to breed. The color has low percentage to get fade, usually kept to enrich the color of the aquarium.

花色类别 / Color ●

红白 / Red White

品种类别 / Species ●

文形 ＞ 背峰变异类 ＞ 琉金型
Fantail Goldfish ＞ Hump Variation ＞ Ryukin

品种详述 / Species Description ●

红白短尾琉金
Red White Ryukin Short Tail

　　红白短尾琉金是短尾琉金中最受欢迎的一个花色品种，虽然只有红白两种花色，但搭配千变万化，绚丽多姿。图例所示6尾短尾琉金体形均属上乘，花色也是入品的花色类型。图例1为白胜更纱型的通背红，一道红色从头至尾，一气呵成，贯穿始终，有气贯长虹之意；图例2为赤胜更纱类型，红色部分殷红如血，白色部分如无瑕美玉，红白对比分明，分界清晰；图例4为六鳞红，因为比传统的十二红多出两抹腮红，因此借用六鳞地金的叫法称为六鳞红；图例5为首尾红，头部为齐鳃红，尾部为全红尾，十分难得。

Red White Ryukin Short Tail is the most popular variety among Ryukin Short Tail. Its color is either red or white, but their combinations are changing and colorful. The one in figure 6 is of top quality of morphology with also a high ranking color. In figure 1, it is Red Back of Shirogachisarasa. The redness extends from head to tail, with the meaning of being full of noble aspiration. In figure 2, it is an Akagachisarasa type goldfish. The red part is bloody and the white part is as white as polished jade, giving a strong contrast view. In figure 4, it is a Rokurin. With two more red gills than the traditional Twelve Red, it is named Rokurin after Rokurin Jikin. In figure 5, it is a Red Head and Tail with red gill and red tail. This is rare to see.

红黑短尾琉金

Red Black Ryukin Short Tail

- **花色类别** / Color

  红黑 / Red Black

- **品种类别** / Species

  文形 > 背峰变异类 > 琉金型

  Fantail Goldfish > Hump Variation > Ryukin

- **品种详述** / Species Description

　　图例所示为红黑短尾琉金，和所有红黑花色一样，红黑琉短尾金也是向红琉金转色的一个过渡期品种。图列1所示红黑短尾琉金较为标准，体型发育完美，背峰高耸，头身比例较为恰当。图例2、3为红黑琉金亚成体，其品种特征尚未充分显现，尾鳍在短尾类型中偏大，而划归宽尾则显得过小，因此品质较之图例1要略逊一筹。和红琉金一样，红黑短尾琉金较易饲养，是适合新手入门学习养殖金鱼的一个品种类型。

The figures show Red Black Ryukin Short Tail. Like all the other red-and-black varieties, it is also a transitional variety from Red Black Ryukin Short Tail to Red Ryukin Short Tail. In figure 1, the Red Black Ryukin Short Tail is perfect with a well-developed body and a proportion with the head. While the ones in figure 2 and figure 3, they are sub-adult which is not fully developed, the caudal fin is a bit larger in Short Tail Variety, but is too small to be classified into Broad Tail, therefore the quality is not as good as the one in figure 1. Like Red Ryukin, Red Black Ryukin Short Tail is easy to breed and is suitable for the fresh man.

黑白短尾琉金

Black White Ryukin Short Tail

- **花色类别** / Color

  黑白 / Black White

- **品种类别** / Species

  文形 > 背峰变异类 > 琉金型

  Fantail Goldfish > Hump Variation > Ryukin

- **品种详述** / Species Description

　　图例所示为黑白短尾琉金亚成体，是蓝琉金中褪色的个体。虽是亚成体，但其背峰发育初具雄姿，体形也较圆润，唯独头和尾的比例略微偏大，可能为幼鱼期饲养密度偏高所致。此黑白色为蓝色褪出，但黑色部分不够浓郁，白色部分质地也不够干净，因此色彩部分略有瑕疵。黑白鱼养殖应注意控制温度，宜偏硬的弱碱性水质为佳，同时要保持水质清新，这样其颜色才能表现完美。

The figure shows a sub-adult of Black White Ryukin Short Tail, and a discolored individual of Blue Ryukin. Although it is a sub-adult, its back hump develops to be initially majestic and its shape is relatively round; only its head and tail occupy a slightly large proportion, which may be caused by relatively high breeding density in the young stage. Such black-and-white color is from blue, but the black part is not heavy enough and the white part is not clean enough, so the colors are slightly defective. As for breeding the Black White Goldfish, the feeder should pay attention to control the temperature, and breed them in slight hard weak alkali water, and keep the water clean, in this way, the color will be perfect.

紫蓝短尾琉金

Chocolate Blue
Ryukin Short Tail

花色类别 / Color •

紫蓝 / Chocolate Blue

品种类别 / Species •

文形 > 背峰变异类 > 琉金型
Fantail Goldfish > Hump Variation > Ryukin

品种详述 / Species Description •

　　图例所示为紫蓝短尾琉金，同雪青短尾琉金类似，也是从蓝短尾琉金中选育出的一个新的花色品种。其基色为青蓝色，身上有铁锈色花斑，显得较为古朴，紫蓝短尾琉金属于小众花色品种。图例所示紫蓝短尾琉金，背峰发育略有欠缺，后期若饲养得法应可以弥补；鳞片细腻有光泽，背鳍高耸有力，可以和其他花色琉金混养。

The figures show Chocolate Blue Ryukin Short Tail. Similar to Lilac Ryukin Short Tail, it is also a new variety developed from Blue Ryukin Short Tail. Its primary color is indigo blue, and on the body, there are rust brown spots, looking primitive and simple. Chocolate Blue Ryukin Short Tail is a minor variety. The one shown in the figure is slightly defective in the back hump development; but if it is well bred in later period, it can be made up. The scales are fine and glossy; the dorsal fin is high and strong; and it can be bred together with other varieties of Ryukin.

三色短尾琉金
Red Black White Ryukin Short Tail

- **花色类别** / Color
  三色 / Red Black White

- **品种类别** / Species
  文形 > 背峰变异类 > 琉金型
  Fantail Goldfish > Hump Variation > Ryukin

- **品种详述** / Species Description

　　图例所示为三色短尾琉金，三色花色多不稳定，黑色部分易褪为红色而成为红白色。图例所示其鳞片黑色部分呈现为黛斑鳞，且黑色浮在鳞片表面，多为繁殖季节由于金鱼性激素作用而产生的一种泛色现象，而繁殖期过后便会消失。图例所示两尾三色短尾琉金体态均属优秀，尤其图例2，背峰背鳍均无瑕疵，有较好的发展空间。

The figures show Red Black White Ryukin Short Tail whose color is generally not stable. The black part is easy to turn red or red white. The black scale shown in the figure is called "Daiban" scale (black scale) where the black color is superficial on the scale, which is a kind of color phenomenon caused by the sex hormone during mating season. After mating season, it will disappear. The two Red Black White Ryukin Short Tail shown in the figures are good in posture, especially the one shown in figure 2 which has perfect back and dorsal fin, indicating great potential.

樱花短尾琉金 — Sakura Ryukin Short Tail (Red White Matt)

- **花色类别** / Color
  软鳞红白 / Red White Matt

- **品种类别** / Species
  文形 > 背峰变异类 > 琉金型
  Fantail Goldfish > Hump Variation > Ryukin

- **品种详述** / Species Description

　　樱花短尾琉金属软鳞红白色，是由五花短尾琉金中分离出来的一个花色品种。软鳞的红色相较硬鳞而显得更为婉约，如寿山芙蓉石般细腻温润。图例所示樱花短尾琉金，体形优异，头尖体圆，背峰高耸，尾鳍短小有力；色彩略有欠缺，红色斑块比例过大，红色饱和度不够，若硬质反光鳞再多3～4枚则恰到好处。

Sakura Ryukin Short Tail (Red White Mat) is a Softs-cale Red White Ryukin and is a variety separated from Calico Ryukin Short Tail. The redness of Soft-Scale seems to be more graceful than that of Hard-Scale. The Sakura Ryukin Short Tail shown in the figure has an excellent body shape, sharp head, round body, high back hump and short but strong caudal fin. It is slightly defective in color. The proportion of red spots is too large and the red saturation is not enough. If there are 3 or 4 more reflective scales, it will be perfect.

云石短尾琉金 | Marble Ryukin Short Tail

- **花色类别** / Color
  软鳞黑白 / Marble

- **品种类别** / Species
  文形 > 背峰变异类 > 琉金型
  Fantail Goldfish > Hump Variation > Ryukin

- **品种详述** / Species Description

　　云石短尾琉金是近几年来新培育的短尾琉金经典品种，是从五花短尾琉金中选育出的软鳞黑白花色。云石花色和水墨花色比较接近，其区别在于云石花色为白多黑少，类似于红白中的白胜更纱。云石的名称由中国水墨山水画中来，其花纹如同茫茫云海中，隐约呈现的一点点山头。云石花色虽为软鳞鱼，但身上有一些硬质反光鳞，在灯光下欣赏有非常炫目的效果。软鳞黑白色不同于硬鳞黑白色，是比较稳定的颜色，通常不会褪色。图例2、3所示云石短尾琉金无论色彩还是体形都恰到好处，是非常优秀的短尾琉金。

Marble Ryukin Short Tail is a new classic variety of Ryukin Short Tail developed in recent years, and it is a Soft-Scale black white variety developed from Calico Ryukin Short Tail. Marble Goldfish is close to the Water Pattern Goldfish. The difference between them is that Marble Goldfish has more white elements than black elements, and is similar to Shirogachisarasa. The name of marble is from Chinese ink landscape paintings because the patterns are like vague hilltops in the sea of clouds. Although Marble Goldfish is a Soft-Scale fish, it has some hard reflective scales on the body, which has a very dazzling effect in the light. Soft-sale Black and White color, different from Hard-Scale Black and White color, is a relatively stable. The Marble Ryukin Short Tail shown in the figure 2 and 3 is perfect in both color and body shape, so it is a very excellent representative variety of Ryukin Short Tail.

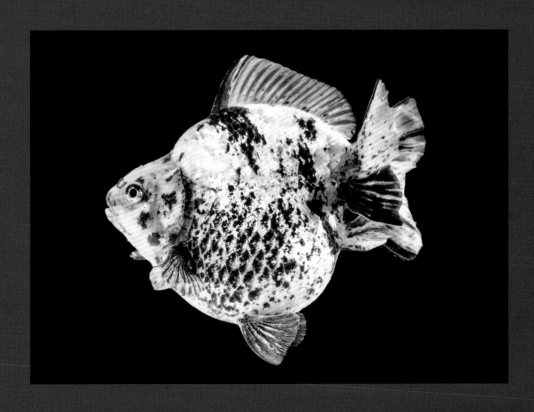

虎纹短尾琉金

Tiger Banded Ryukin Short Tail

- **花色类别** / Color
  软鳞红黑 / Tiger Banded

- **品种类别** / Species
  文形 > 背峰变异类 > 琉金型
  Fantail Goldfish > Hump Variation > Ryukin

- **品种详述** / Species Description

　　图例所示为虎纹短尾琉金。从外部形态上来看，图例所示均属上乘；而从色彩上说，图例1的底色要胜过图例2、3，图例2、3底色偏黄，而图例1为较好的橙红色，若黑色条纹再多2～3条则更完美。虎纹短尾琉金是近些年来才流行的一个花色类型，但对底色要求严格，好的虎纹花色也不容易获得。

The figures show Tiger Banded Ryukin Short Tail. In terms of external form, Tiger Banded Ryukin Short Tail in the figures are of top quality. In terms of color, the bottom color of the one in figure 1 is better than that of those in figure 2 and 3, in which the bottom color is a little yellow. In figure 1, orange red is better presented, and it will be more perfect, if there are 2-3 more black stripes. Tiger Banded Ryukin Short Tail is a variety which has become popular in recent years, but the requirement for bottom color is strict, so it is not easy to find a good one.

麒麟短尾琉金

Kirin Ryukin
Short Tail

- **花色类别** ∕ Color
  软鳞五花∕ Kirin

- **品种类别** ∕ Species
  文形 ＞ 背峰变异类 ＞ 琉金型
  Fantail Goldfish ＞ Hump Variation ＞ Ryukin

- **品种详述** ∕ Species Description

　　图例所示是麒麟短尾琉金。麒麟色是软鳞五花中出现全身青灰色或青黑色的花色类型，其鳞片一般基部颜色较深，边缘处较浅，排列整齐，十分富有层次感。纯青灰色麒麟花色为墨麒麟，而身上带色斑的称花麒麟。麒麟花色如果配以红顶者，会显得更加高贵。麒麟是传说中的瑞兽，因此现在比较受欢迎。

The figure shows Kirin Ryukin Short Tail; Kirin Color is gray green or green black among Soft-Scale Calico (Bluish Base) Goldfish. Generally, the primary color of its scales is heavy while the color of the edges is light, and the scales are neat and layered. The pure cinereous Kirin Goldfish are known as Black Kirin Goldfish while those with spots are known as Calico Kirin. Kirin Goldfish growing with a red head pompon is regarded nobler. It is quite popular now since Kirin is a legendary auspicious beast.

花短尾琉金

Calico Ryukin
Short Tail

- **花色类别** / Color
  软鳞五花 / Calico

- **品种类别** / Species
  文形 > 背峰变异类 > 琉金型
  Fantail Goldfish > Hump Variation > Ryukin

- **品种详述** / Species Description

　　花短尾琉金是五花琉金中选育出的花色品种，其外形特征和其他短尾琉金相同。图例1所示花短尾琉金背峰凸起稍有欠缺，但花色美艳，白色基底上流动嫣红色块，与鱼鳍上黑丝纹相得益彰，相互映衬。图例2、3的花琉金，其体型较之图例1要更加标准。花短尾琉金没有完全一样的花色，千变万化，因而饲养一群或配以其它花色都可以取得很好的观赏效果。

Calico Ryukin Short Tail is a variety developed from Calico Ryukin, and its appearance features are the same as other short-tail Ryukin. Calico Ryukin Short Tail shown in the figure is defective in back hump bulge, but its color is beautiful and gorgeous. The bright red spots on the white base complement the black marks on the fins. The body of the ones in figure 2 and 3 are better than figure 1. Calico Ryukin Short Tail is changing in color, so a school of them or a mixture with other colors of goldfish can have a good appreciation effect.

重彩短尾琉金 — Ryukin Short Tail (with Intense Colors)

- **花色类别** / Color

  软鳞五花 / Intense Colors

- **品种类别** / Species

  文形 ＞ 背峰变异类 ＞ 琉金型

  Fantail Goldfish ＞ Hump Variation ＞ Ryukin

- **品种详述** / Species Description

  重彩短尾琉金，是近些年来选育的一个新的花色品种，也广受市场好评。花色创新上采用重彩，也是对中国金鱼传统色彩审美的一种突破，软鳞的基底加上厚重的色调，如同传统重彩水墨画一般，对欣赏者的视觉冲击力十分强。其雄健的体形，配以浓墨重彩的颜色，好似喋血沙场的武士，"龙绦金盔镶铁甲，百战敌血染征袍"，充满着阳刚之美。重彩短尾琉金特别适合玻璃水族箱观赏，配以适合的顶光，可以将其如同李可染先生山水画代表作《万山红遍》的诗意效果，完美地诠释出来。

  Ryukin Short Tail (with Intense Colors) is a new variety developed in recent years, and is widely praised in the market. In color innovation, heavy color is used, which is a breakthrough in the traditional color aesthetics. The Soft-Scale base and heavy colors, like the traditional heavy-color ink and wash paintings, has a strong visual impact for the appreciators. Its strong body shape and heavy colors make it full of masculine beauty like a warrior in the battlefield. Ryukin Short Tail (with Intense Colors) is particularly suitable for appreciation in glass aquarium. In the proper top light, it can perfectly interpret the poetic effect of Mr. Li Keran's representative landscape painting "Mountains Are Reddened".

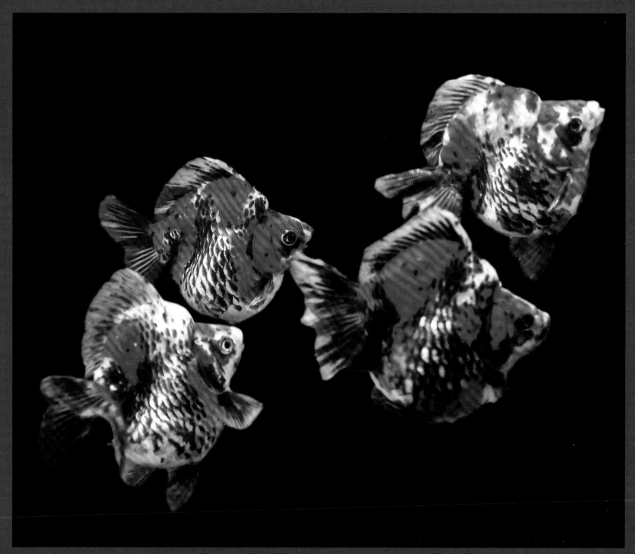

五花短尾琉金

Calico Ryukin Short Tail (Bluish Base)

- **花色类别** / Color
  软鳞五花 / Calico (Bluish Base)

- **品种类别** / Species
  文形 > 背峰变异类 > 琉金型
  Fantail Goldfish > Hump Variation > Ryukin

- **品种详述** / Species Description

　　五花短尾琉金是短尾琉金中较早出现的一种花色品种，也一直是市场中颇受欢迎的一个花色品种。短尾琉金体如圆盘，可以有足够的空间，很好地展现出五花色的丰富层次和软鳞花色如大理石般的肌理效果。图例1所示五花短尾琉金体型优美，淡蓝色背脊逐渐过渡到白色腹部，如同天青色汝窑瓷器般的典雅高贵，点缀的白色硬质反光鳞和黑色碎花恰到好处，红面玉顶又显俏皮。

Calico Ryukin Short Tail (Bluish Base) is a variety appearing earlier among the Ryukin Short Tail, and is also very popular in the market. Ryukin Short Tail has a disc-like body, so it has enough space to present the rich layers of Calico goldfish and marble-like texture effect of Soft-Scale goldfish. Calico Ryukin Short Tail (Bluish Base) shown in figure 1 has a beautiful body shape and its light blue back gradually transits to the white belly, looking elegant and noble; The white reflective scale and black patches are perfect; and red face and jade-like top make it cuter.

软鳞五花 / Calico (Bluish Base)

红宽尾琉金
Red Ryukin Broad Tail

- **花色类别** / Color

  红 / Red

- **品种类别** / Species

  文形 > 背峰变异类 > 琉金型

  Fantail Goldfish > Hump Variation > Ryukin

- **品种详述** / Species Description

　　红宽尾琉金，是宽尾琉金中最常见的一个花色品种。纯红色的金鱼或红白色的金鱼有时会在鱼鳍、背部、腹部或头部出现黑色的条纹或斑块，而过段时间又会消失，有可能是水质变化造成。繁殖季也会出现这类情况，多为其激素分泌造成的泛色现象。一般不需要特别治疗。图例所示红宽尾琉金较为优秀，体型饱满圆润，背峰和背鳍高耸，尾鳍飘逸如翩翩舞裙，在同一尾鱼的不同视角对比下，我们可以知道宽尾琉金是非常适合侧视欣赏的一个品种。

The figure shows Red Ryukin Broad Tail, which is the commonest variety of Ryukin Broad Tail type. Black stripes or plaques sometimes will appear on the fins, backs, bellies or heads of the purely red goldfish or red white goldfish, and disappear after a while, which may be caused by water quality change. This will also happen in the breeding seasons, and this is mainly the color flooding phenomenon caused by hormone excretion. Usually, special treatment is not needed. The one in the figure has excellent body, high back hump and dorsal fin. The wide stretching caudal fin is like the graceful dress. Compared with other varieties, Red Ryukin Broad Tail is better to view from sides.

白宽尾琉金

White Ryukin Broad Tail

- **花色类别** / Color

  白 / White

- **品种类别** / Species

  文形 > 背峰变异类 > 琉金型

  Fantail Goldfish > Hump Variation > Ryukin

- **品种详述** / Species Description

  白色在宽尾长鳍的各种文形金鱼中会有很好的表现，图例所示的白宽尾琉金就是非常好的印证。此鱼背峰高耸，体态雄健，尾鳍宽大飘逸，外形堪称完美。全身洁白，如一身高贵的婚纱，而胸鳍的第一鳍梗、眼睛和嘴唇上，恰到好处地点缀上一抹红色，妩媚却不显妖娆。

  White is well presented in various Fantail Goldfish with long and wide caudal fins, and White Ryukin Broad Tail shown in the figure well confirms it. This goldfish has a high back hump, vigorous posture and a wide and elegant caudal fin, so the appearance is really perfect. It is white in the whole body, like a noble wedding dress; some redness is dotted on the first fin stem of pectoral fin, eyes and lips, making it charming but not enchanting.

红白宽尾琉金

Red White Ryukin Broad Tail

- **花色类别** / Color
  红白 / Red White

- **品种类别** / Species
  文形 > 背峰变异类 > 琉金型
  Fantail Goldfish > Hump Variation > Ryukin

- **品种详述** / Species Description

　　图例所示为红白宽尾琉金。和大多数红白花色金鱼一样，红白宽尾琉金也是宽尾琉金中最受欢迎的一个花色品种，虽然只有红白两种花色，但搭配千变万化，绚丽多姿。图例1所示为赤胜更纱；图例2、3为通背红，两尾宽尾琉金体形均属上乘，若赤斑鳞花色是清丽俏皮，那么这两尾则如贵妇般雍容端庄。

The figures show Red White Ryukin Broad Tail. Like most of the red white goldfishes, Red White Ryukin Broad Tail is also one of the most popular varieties among the Ryukin Broad Tail. They are red or white, but the combinations are changing and colorful. Figure 1 presents an Akagachisarasa; figure 2 and 3 present Red Back, both of which are of high quality in body shape. If Red White Ryukin Broad Tail (with Red Dots) is regarded as elegant and beautiful, then the two here are more graceful and dignified.

赤斑鳞宽尾琉金
Red White Ryukin Broad Tail (with Red Dots)

- **花色类别** / Color
  红白 / Red White (with Red Dots)

- **品种类别** / Species
  文形 > 背峰变异类 > 琉金型
  Fantail Goldfish > Hump Variation > Ryukin

- **品种详述** / Species Description

　　图例所示为赤斑鳞宽尾琉金。赤斑鳞又称鹿子花，是在白色的鳞片中点缀点点红色斑鳞的一种花色。赤斑鳞是红白色中培育出的花色品种。图例4所示是通背红配以赤斑鳞，花色艳美，背峰及背鳍高耸，头尖体圆，身形饱满，尾鳍舒展飘逸，犹如水中盛开的牡丹。图例1、2、3全身满布赤斑鳞，配以一抹红唇，十分俏皮可爱。

The figures show Red White Ryukin Broad Tail (with Red Dots). Red Dots Scale is also called Sika Deer. It is a color of white scale dotting with red spot scales. Red Dots Scale Goldfish is a variety developed from Red White Goldfish. In figure 4, Red Back and Red Dots Scale Goldfish are shown together. With beautiful color, high back hump and dorsal fin, sharp head, round and plump body, extended and graceful caudal fin, it looks like blooming peony. The goldfish in figure 1, 2 and 3 are covered with red dots scales, and with the red lips, they look cute and lovely.

红黑宽尾琉金

Red Black Ryukin Broad Tail

- **花色类别** / Color

  红黑 / Red Black

- **品种类别** / Species

  文形 > 背峰变异类 > 琉金型

  Fantail Goldfish > Hump Variation > Ryukin

- **品种详述** / Species Description

  　　图例所示为红黑宽尾琉金。和所有红黑花色一样，红黑宽尾琉金也是向红宽尾琉金转色的一个过渡期品种。图例所示红黑宽尾琉金背峰及背鳍高耸，头尖体圆，身形饱满，尾鳍舒展飘逸。红黑宽尾琉金较易饲养，是适合新手入门学习养殖金鱼的一个品种类型。宽尾琉金特别适合大型水族箱养殖，侧视观赏效果最佳。

  The figure shows Red Black Ryukin Broad Tail. Like all other Red Black Goldfish, it is also a transitional variety from Black Ryukin Broad Tail to Red Ryukin Broad Tail. The Red Black Ryukin Broad Tail shown in the figure has a high back hump and dorsal fin, sharp head, round and plump body, extended and graceful caudal fin. It is easy to breed, so it is a variety suitable for the green hands. And it is particularly suitable for breeding in large aquarium, and the appreciation effect from sides is the best.

黑白宽尾琉金

Black White Ryukin
Broad Tail

- **花色类别** / Color

  黑白 / Black White

- **品种类别** / Species

  文形 > 背峰变异类 > 琉金型

  Fantail Goldfish > Hump Variation > Ryukin

- **品种详述** / Species Description

  　　图例所示为黑白宽尾琉金。黑白宽尾琉金是从蓝宽尾琉金中选育而出，通常是蓝宽尾琉金褪色时期过渡阶段的一个品种，一般很难达到比较稳定的黑白色。但是，我们可以通过人为地控制水质、温度以及在饵料中添加螺旋藻等方法延缓褪色的过程。图例2、3近似熊猫花色，黑背白腹，黑眼黑鳍。图例1是黑白花中的巧色玉面，全身墨黑唯独头面洁白，黑目黑唇，很是难得。

  The figures show Black White Ryukin Broad Tail, which is selected and bred from Blue Ryukin Broad Tail. Usually, it is a transitional variety when the color of Blue Ryukin Broad Tail fades, and generally, it is hard to reach a relatively stable Black White. However, we can delay the color fading process through artificially controlling the water quality, temperature and adding spirulina to the bait. The goldfish in figure 2 and 3 are similar to the color of panda, with black back, white belly, black eyes and fins. The goldfish in figure 1 is ingenious color with Jade face of Black White Goldfish. It is entirely black except the white head, and has black eyes and lips, which is very rare.

花宽尾琉金

Calico Ryukin Broad Tail

- **花色类别** / Color

  软鳞五花 / Calico

- **品种类别** / Species

  文形 > 背峰变异类 > 琉金型

  Fantail Goldfish > Hump Variation > Ryukin

- **品种详述** / Species Description

  图例所示为花宽尾琉金，是从五花宽尾琉金中选育出的一个花色品种。此尾花宽尾琉金颜色较素雅，反光鳞点缀较为适度。图例1中所示外形背峰较好，但背鳍先端有轻微倒鳍的迹象。尾鳍宽大飘逸，婀娜多姿，十分适宜水族箱侧视玩赏。图例3为红顶花色，特别受到玩赏者的青睐。图例5、6、7为荧鳞花色褪色出的花宽尾琉金，鳞片反光效果好，但尾鳍略微偏小。

  The figures show Calico Ryukin Broad Tail, a Soft-scale tri-colored goldfish, which is selected and bred from Calico Ryukin Broad Tail (Bluish Base). In figure1, it has a fine back hump though the front edge of the fin is a little inclined. But with a broad and stretching fin, it is suitable to view from side in aquarium. The one in figure 3 has red head pompon, which is favored by goldfish fanciers. The ones in figure 5, 6, and 7 are Calico Ryukin Broad Tail transited from Fluorescent Scale with good scale reflection, but the caudal fin is a little bit smaller.

软鳞白宽尾琉金

White Matt Ryukin
Broad Tail

- **花色类别** / Color
  白 / White Matt

- **品种类别** / Species
  文形 > 背峰变异类 > 琉金型
  Fantail Goldfish > Hump Variation > Ryukin

- **品种详述** / Species Description

　　图例所示为软鳞白宽尾琉金，为软鳞五花中选育出的花色品种。其全身若羊脂白玉，温润柔白；鱼鳍基部和头部的淡柠檬黄色又如包裹的沙金玉皮，巧然天成；鱼体丰满，背峰及背鳍高耸，尾鳍宽大飘逸，游动时清丽脱俗。宽尾琉金体质强健，易于饲养。玩赏以大型水族箱辅以灯光，其侧视最佳。

The figure shows White Matt Ryukin Broad Tail, which is a variety developed from Soft-Scale Calico (Bluish Base) Goldfish. Its whole body is as white as jade. The light lemon yellow on the base of fins and the head is like golden skin. It has a plump body, high back pump and dorsal fin, broad and graceful caudal fin, so it looks beautiful and refined when swimming. White Matt Ryukin Broad Tail is strong and easy to breed. It is better to appreciate them in a large aquarium with light from sides.

花色类别 / Color •

软鳞红白 / Red White Matt

品种类别 / Species •

文形 > 背峰变异类 > 琉金型
Fantail Goldfish > Hump Variation > Ryukin

品种详述 / Species Description •

樱花宽尾琉金
Sakura Ryukin Broad Tail
(Red White Matt)

　　图例所示为樱花宽尾琉金。其背峰及背鳍高耸，头尖体圆，身形饱满，尾鳍舒展飘逸。樱花花色为软鳞，全身若白玉羊脂，温润柔白，配以碎红花，有如点点春桃，又如寿山芙蓉石细腻如凝脂。樱花宽尾琉金型色皆优的非常珍稀，图例所示之鱼虽为亚成体，但饲养得当将来会成为精品金鱼。

The figure shows Sakura Ryukin Broad Tail (Red White Matt), with a high back hump and dorsal fin, a sharp head and a round and plump body, extended and graceful caudal fin. The Sakura color refers to the color of the Soft-Scale. Its whole body is white dotted with breaking redness, just like a fresh pink peach or the lotus marble in Shoushan Mountain. Sakura Ryukin Broad Tail (Red White Matt) that is excellent in both shape and color is very rare. Although the goldfish in figure is a sub-adult, it may become a high-quality goldfish if it is well fed.

水墨宽尾琉金
Water Pattern
Ryukin Broad Tail

- 花色类别 / Color
  软鳞黑白 / Water Pattern

- 品种类别 / Species
  文形 > 背峰变异类 > 琉金型
  Fantail Goldfish > Hump Variation > Ryukin

- 品种详述 / Species Description

　　水墨宽尾琉金也是近几年刚刚诞生的一个新的花色品种，是金鱼业者从五花宽尾琉金中选育出的。云石花色和水墨花色比较接近，其区别在于水墨花色为黑多白少，类似于红白中的赤胜更纱。水墨宽尾琉金由于是软鳞，其花色表现效果如同在生宣纸上水墨晕染的效果，又像大理石的肌理效果，让人感叹造化的神奇。水墨宽尾琉金鳞片上消耗的钙磷等较硬鳞鱼少，可以让骨骼充分发育，所以一般软鳞花色的鱼都较硬鳞花色的鱼体形大。

Water Pattern Ryukin Broad Tail is a new variety selected and bred from Calico Ryukin Broad Tail (Bluish Base) in recent years. The Marble color is close to the Water Pattern color. Their difference lies in that the latter has more black elements than white elements and is similar to Akagachisarasa of Red White ones. Because Water Pattern Ryukin Broad Tail is a Soft-Scale goldfish, its color gives an effect of ink blooming on untreated Xuan paper and also a textural effect of marble, which is really amazing. Calcium and phosphorus consumed on the scales of Water Pattern Ryukin Broad Tail are less than the Hard-Scale goldfish, so the bones can fully develop. Therefore, generally, the Soft-Scale goldfishes are larger than the Hard-Scale ones.

花色类别 / Color •

软鳞五花 / Bluish Matt and Black Dots

品种类别 / Species •

文形 > 背峰变异类 > 琉金型

Fantail Goldfish > Hump Variation > Ryukin

品种详述 / Species Description •

撒锦宽尾琉金 ‖ Ryukin Broad Tail (with Bluish Matt and Black Dots)

撒锦宽尾琉金为软鳞花色，和云石花色、水墨花色一样，由软鳞五花选育而出。其特点为白地如雪地，点点墨色点缀其上，又如在宣纸上撒上金片碎粒，故称之为撒锦，俗称芝麻花。与云石的差别是，云石的墨呈条纹状分布，而撒锦大部分墨呈现点状分布。软鳞黑白系列颜色较硬鳞黑白系列稳定，是宽尾琉金黑白色系发展的一个大趋势。

Ryukin Broad Tail (with Bluish Matt and Black Dots) is a Soft-Scale Calico (Bluish Base) Goldfish, and like Marble color and Water Pattern color, it is also selected and bred from Matt Calico Goldfish. It is featured by snow white base with dotted blackness, like gold pieces on Xuan paper (professional paper for water pattern art). The difference between it and a Marble one is that: the blackness on a Marble one usually forms stripes while that on a Ryukin Broad Tail (with Bluish Matt and Black Dots) usually forms dots. The color of Soft-Scale black white series is more stable than that of the Hard-Scale black white series, so it is a great development trend of Black White Ryukin Broad Tail.

麒麟宽尾琉金

Kirin Ryukin Broad Tail

- **花色类别** / Color

  软鳞五花 / Kirin

- **品种类别** / Species

  文形 > 背峰变异类 > 琉金型

  Fantail Goldfish > Hump Variation > Ryukin

- **品种详述** / Species Description

  图例所示为麒麟宽尾琉金。麒麟宽尾琉金由五花宽尾琉金选育而出，全身青黑色软鳞，鳞片基部色较深，边缘略浅，排列整齐，很有层次感。图例2为红顶花麒麟色。麒麟宽尾琉金一直是市场认同度颇高的一个宽尾琉金花色品种。

  The figures show Kirin Ryukin Broad Tail, which is selected and bred from Calico Ryukin Broad Tail. Its body is covered with blue black Soft-Scale; the color in base of the scales is heavier than that in the edges, and the scales are neat and layered. In figure 2, it is a Red Cap Kirin Ryukin Broad Tail. Kirin Ryukin Broad Tail has always been a Ryukin Broad Tail variety winning good market reactions.

重彩宽尾琉金
Ryukin Broad Tail (with Intense Colors)

- **花色类别** / Color
  软鳞五花 / Intense Colors

- **品种类别** / Species
  文形 > 背峰变异类 > 琉金型
  Fantail Goldfish > Hump Variation > Ryukin

- **品种详述** / Species Description

　　重彩宽尾琉金为软鳞花色，由麒麟花色选育出，是近些年来最受市场青睐的一个花色品种。软鳞花色中的黑色部分称为墨，红色部分称为绯，白色部分称为地。重彩讲究浓绯重墨，下腹部可以出地，侧线起至背脊部分，显现的地越少越好。图例2所示金鱼，体形饱满圆润，背峰及背鳍高耸，尾鳍舒展有力，唯脊背部露地过多，绯色太少。重彩花色在水族缸中一定要辅以灯光观赏，才能将其绚烂之美表现得淋漓尽致。

Ryukin Broad Tail (with Intense Colors), a Soft-Scale goldfish, is selected and bred from Kirin ones; and it is one of the most popular varieties in the market in recent years. The black part on Ryukin Broad Tail (with Intense Colors) is called Mo (ink); the red part is called Fei (dark red); the white part is called Di (ground white). For the appreciators, Ryukin Broad Tail (with Intense Colors) should have more dark red parts and ink parts, and there may be ground whiteness in the belly, but there should be as little ground whiteness as possible in the part from siding to back. The goldfish in figure 2 is plump and round in body, has a high back hump and dorsal fin, and extended and strong tail fins, but there is too much the ground whiteness on the back and too little dark redness. The beauty of splendor of Ryukin Broad Tail (with Intense Colors) can be well shown with light in the aquarium.

紫长尾琉金 | Chocolate Ryukin Long Tail

- **花色类别** / Color

  紫 / Chocolate

- **品种类别** / Species

  文形 > 背峰变异类 > 琉金型

  Fantail Goldfish > Hump Variation > Ryukin

- **品种详述** / Species Description

　　紫长尾琉金雅称紫凤凰，由于其尾甚长，超过身体3倍，尾叶又窄，酷似凤凰飘带状尾羽，故此得名。相较于短尾琉金和宽尾琉金而言，长尾琉金现在饲养的鱼场越来越少。图例所示紫长尾琉金虽体形娇小，倒也俏皮可人，头部比例略微偏大，其尾鳍飘逸舒展，配以古铜色的体色，给人以清雅之美。

Chocolate Ryukin Long Tail is known as Chocolate Phoenix, because its tail is very long, over three times as long as its body, and its tail leaf is narrow, exactly like the streamer-like tail feather of phoenix. Compared to the Ryukin Short Tail and Ryukin Broad Tail, fewer and fewer fish farms feed Ryukin Long Tail now, mainly in Jiangsu. The Chocolate Ryukin Long Tail in the legend is small, but looks beautiful and cute. The proportion of its head is slightly larger; its caudal fin are graceful and extended; coupled with the bronze body color, it looks elegant.

红白琉金 — Red White Ryukin

- **花色类别** / Color
  红白 / Red White

- **品种类别** / Species
  文形 > 背峰变异类 > 琉金型
  Fantail Goldfish > Hump Variation > Ryukin

- **品种详述** / Species Description

　　图例所示为红白琉金，是较常见的一个琉金花色品种，也是一个基础品种，长尾、短尾和宽尾的都是通过它选育出来的。红白琉金适合大面积的坑塘饲养，且需要一定的深度，适合水族箱侧视玩赏。红白琉金体质雄健，适应力强，适合群养，可以和高身龙睛、宽尾琉金等混养，但不宜与水泡、珍珠等混养。

The figures show Red White Ryukin, which is a relatively common variety of Ryukin and also a basic variety. The long-tail, short-tail and wide-tails are all developed from it after selection. Red White Ryukin is suitable to breed in large-area ponds with a certain depth; it is suitable for appreciation from sides in aquarium. Red White Ryukin is strong and highly adaptable, and is suitable for group breeding. It can be bred together with Telescope with High Hump and Ryukin Broad Tail, but should not be bred together with Bubble-Eye and Pearlscale.

花色类别 / Color •
软鳞红白 / Red White Matt

品种类别 / Species •
文形 > 背峰变异类 > 琉金型
Fantail Goldfish > Hump Variation > Ryukin

品种详述 / Species Description •

樱花琉金
Sakura Ryukin
(Red White Matt)

　　图例所示为樱花琉金，樱花为软鳞红白色。此鱼体形较佳，而色彩有所欠缺。其绯的色彩饱和度不够，偏橙红色，而绯与地的分布比例尚可，但绯的位置如果在背脊处则最佳，或满布下腹部成为卷腹红；硬质反光鳞略有偏上，能达到4～6个则最佳。樱花琉金在一缸重色琉金之中可以起到很好的配色效果，适宜水族箱辅以灯光玩赏。

The figures show Sakura Ryukin (Red White Matt), and Sakura refers to Soft-Scale Red White. This goldfish has a good body shape, but its colors are not very good. The color saturation of its dark redness is not enough, more like orange red; the distribution proportion of redness and whiteness is acceptable, but if the redness is on the back or covers the whole belly, it will be the best. If the hard reflective scales are a little upward and the number reaches 4-6, it will be the best. Sakura Ryukin (Red White Matt) can play a good matching effect in an aquarium of Ryukin Broad Tail (with Intense Colors), and it is suitable for appreciation in the aquarium with light.

五花琉金 Calico Ryukin (Bluish Base)

- **花色类别** / Color
  软鳞五花 / Calico (Bluish Base)

- **品种类别** / Species
  文形 > 背峰变异类 > 琉金型
  Fantail Goldfish > Hump Variation > Ryukin

- **品种详述** / Species Description

　　图例所示为五花琉金。其外部形态较好，背峰高耸，背脊宽厚，体形侧视如圆盘状，尾鳍属于中尾类型，背鳍先端不够挺拔，有轻微倒鳍的迹象，是较明显的瑕疵；颜色上基本属于五花类型，其后部颜色优于前半部颜色，前半部红色部分所占面积过大，且红色的饱和度不够，反光鳞略显偏多。

The figures show Calico Ryukin (Bluish Base) which has good appearance, high and thick back. As viewing from the side, it looks like a round plate. Its caudal fin is middle sized. However, the front end of the dorsal fin is not straight, and tends to incline, which is an obvious defect. It can be classified into five color type, with better color matching at the rear, as in the front, the red area is too large and the color is not saturated, and there is too much reflective scales.

文形
Fantail Goldfish

∨

鳞片变异类
Scale Variation

∨

皮球珍珠型
Golfball Pearlscale

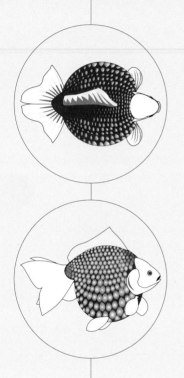

　　皮球珍珠是鳞片变异类的代表品种，它的出现，在中国金鱼发展史上，有着开宗创派的重要意义。关于皮球珍珠还有一段野史，清嘉庆年间，一批中国金鱼传入印度，受环境气候以及水质等因素影响，变异出红皮球珍珠。其鳞片石灰质沉积，中央凸起，边缘色深，中间色浅，如同串串珍珠排列，极富立体美感。后慈禧太后重修圆明园，印度将红皮球珍珠作为礼品进献。但因不适应北京的气候，第一批金鱼全部死亡；在第二批送来后，时任两广总督李鸿章私自留下一部分给了任广州海关总监的女婿任思九。任思九将这些红皮球珍珠私藏在上海的虹口花园，秘不示人。次年，有爱好者买通了他家的佣人，将附在水草上的鱼卵偷了出来，孵出了幼鱼。又过了一年，将这些红皮球珍珠和五彩金鱼杂交，培育出五花皮球珍珠。后来又演变为各色皮球珍珠。

Golfball Pearlscale Goldfish is a representative variety of scale variation, and its appearance has an important significance of starting a family in the development history of Chinese goldfish. There is an unofficial history about Golfball Pearlscale Goldfish. In the reign of Emperor Jiaqing of the Qing Dynasty, some Chinese goldfish were introduced to India, and due to the influence of such factors as environmental climate and water quality, Red Golfball Pearlscale Goldfish appeared through variation. Its scales had deposit of calcium carbonate, are raised in centers; and the color in the edges is deeper than that in the middle; the scales arranged like pearls, presenting a three-dimensional beauty. Later, when Empress Dowager Cixi repaired the Old Summer Palace, India offered Red Golfball Pearlscale Goldfish as a gift. But because they did not adapt to the climate of Beijing, all of them died; after another school was sent, Li Hongzhang, then-Governor of Liangguang privately left some to his son-in-law Ren Siqiu, the then-director of Guangzhou Customs. And Ren Siqiu hid these Red Golfball Pearlscale Goldfish in Hongkou Garden in Shanghai. In the next year, a goldfish lover bought off his servant, stealing the roes on the water plants and hatched juvenile fish. Another year later, these Red Golfball Pearlscale Goldfish and colorful goldfish were bred together and got Calico Golfball Pearlscale (Bluish Base). Then, Golfball Pearlscale Goldfish of different colors appeared.

红皮球珍珠 | Red Golfball Pearlscale

花色类别 / Color •

红 / Red

品种类别 / Species •

文形 > 鳞片变异类 > 皮球珍珠型

Fantail Goldfish > Scale Variation > Golfball Pearlscale

品种详述 / Species Description •

　　图例所示为红皮球珍珠。红皮球珍珠原为印度传入我国，进贡于当年的西太后，其中一批与其他花色的珍珠鳞金鱼杂交，产生了现代各种花色的皮球珍珠。红皮球珍珠以体圆似球，尖嘴凹眼、珍珠鳞饱满者为贵。以前多小尾，现在也有不少大尾的皮球珍珠。红皮球珍珠饲养要求高，抗病力差，体质较弱，不易饲养。红皮球珍珠以俯视玩赏为主，多以木海、砂缸、陶盆等容器饲养。

The figures show Red Golfball Pearlscale, which was originally introduced to China from India and offered to Empress Dowager Cixi. Some of them were bred together with Pearlscale Goldfish of other colors, producing modern Pearlscale Goldfish of various colors. Among the Red Golfball Pearlscale Goldfish, the ones with ball-like round body, sharp mouth, deep-set eyes and full pearscale are precious. In the past, most of them were small-tail ones, but now, there are also many large-tail ones. The requirement for feeding Red Golfball Pearlscale is very high because it has a poor resistance to disease and is weak. Therefore, it is not easy to feed them. Red Golfball Pearlscale Goldfish are mainly appreciated from above, and they are usually fed in such vessels as wooden basin, sand tank and pottery basin.

- **花色类别** / Color
  蓝 / Blue

- **品种类别** / Species
  文形 > 鳞片变异类 > 皮球珍珠型
  Fantail Goldfish > Scale Variation > Golfball Pearlscale

- **品种详述** / Species Description

　　图例所示为蓝皮球珍珠。蓝皮球珍珠属于小众品种，饲养的不多，目前江苏南京、徐州等地鱼场有小量饲养。蓝皮球珍珠全身墨蓝色，全身覆石灰质鳞片，颗粒饱满度适中。蓝皮球珍珠多大尾，遇高温时，有部分会转色为黑白皮球珍珠和三色皮球珍珠，甚是有趣。

The figures show Blue Golfball Pearlscale, which is enjoyed in a minor group. There are not many, and currently only a small quantity are rasied in Nanjing and Xuzhou of Jiangsu Province. Blue Golfball Pearlscale is blue black in the whole body, and it is covered with calcareous scales that are moderately plump. Most of the Blue Golfball Pearlscale Goldfish are large-tail goldfish. In high temperature, some of them will become Black White Golfball Pearlscale and Tri-colored Golfball Pearlscale, which is very interesting.

紫皮球珍珠 | Chocolate Golfball Pearlscale

花色类别 / Color ·
紫 / Chocolate

品种类别 / Species ·
文形 > 鳞片变异类 > 皮球珍珠型
Fantail Goldfish > Scale Variation > Golfball Pearlscale
品种详述 / Species Description ·

　　紫皮球珍珠是由红皮球珍珠选育而出的。紫皮球珍珠全身赤铜色，珍珠鳞饱满度适中，但饲养不好易出平鳞。紫皮球珍珠属于小众花色品种，饲养地方不多，主要在江苏南京、徐州，安徽阜阳，目前天津也有引入饲养的。紫皮球珍珠易出小尾，而小尾的紫皮球珍珠亦显得十分俏皮。

Chocolate Golfball Pearlscale is selected and bred from Red Golfball Pearlscale. Chocolate Golfball Pearlscale is red bronze in the whole body, and its pearlscales are moderately plump, but flat scale may appear easily if it is not well fed. Chocolate Golfball Pearlscale is a minor variety, and is not fed in many places, mainly in Nanjing and Xuzhou in Jiangsu, and Fuyang in Anhui. At present, Tianjin also introduces it. It is easy to feed small-tail Chocolate Golfball Pearlscale, which look very cute.

红白皮球珍珠

Red White Golfball Pearlscale

- **花色类别** / Color
  红白 / Red White

- **品种类别** / Species
  文形 ＞ 鳞片变异类 ＞ 皮球珍珠型
  Fantail Goldfish ＞ Scale Variation ＞ Golfball Pearlscale

- **品种详述** / Species Description

　　图例所示为红白皮球珍珠。红白皮球珍珠是在红皮球珍珠的基础上选育出的红白色个体。1993年，红白花皮球珍珠在江苏无锡举办的中国金鱼展品会上曾获银奖。在红白皮球珍珠中，知名度较高的是由武汉金鱼业者培育的，以通背红皮球珍珠为代表的各色红白皮球珍珠。

The figures show Red White Golfball Pearlscale, which is a red white individual selected and bred from Red Golfball Pearlscale. In 1993, Red White Golfball Pearlscale won the silver prize at the Chinese Goldfish Exhibition held in Wuxi, Jiangsu. Among the Red White Golfball Pearlscale Goldfish, the Red White Golfball Pearlscale of different colors represented by Red Back Golfball Pearlscale fed by Wuhan fish farmers are more famous.

五花皮球珍珠 ｜ Calico Golfball Pearlscale (Bluish Base)

- **花色类别** ／ Color

  珍珠鳞五花 ／ Calico (Bluish Base)

- **品种类别** ／ Species

  文形 > 鳞片变异类 > 皮球珍珠型

  Fantail Goldfish > Scale Variation > Golfball Pearlscale

- **品种详述** ／ Species Description

  图例所示为五花皮球珍珠。五花皮球珍珠是目前皮球珍珠中，最受市场追捧的一个花色品种。上等的五花皮球珍珠尖头凹眼，体圆如球、珍珠鳞饱满；花色上讲究素净，红顶、白肚、蓝背、黑碎花，黑色和红色斑块不宜过多过重。皮球珍珠不耐缺氧，一旦缺氧，眼珠就会外凸，称为爆眼，是饲养皮球珍珠的大忌。

The figures show Calico Golfball Pearlscale (Bluish Base), which is the most popular varieties in the market among all the Golfball Pearlscale goldfishes. The superior Calico Golfball Pearlscale (Bluish Base) has sharp head, deep-set eyes and plump pearlscale. The colors are quiet-red cap, white belly, blue hump, black pieces; there should not be too much and too heavy black and red plaques. The ones produced in Tianjin, Wuhan, Nanjing, Nantong and Suzhou are of higher quality. Golfball Pearlscale is intolerant to oxygen deficit. Once there is a shortage of oxygen, the eyeballs will become convex, which is a taboo of feeding Golfball Pearlscale.

文形
Fantail Goldfish

∨

尾鳍变异类
Tail Fin Variation

∨

文蝶型
Butterfly Tail

　　文蝶是近几年刚推出的一个金鱼新品种。其头身如文鱼前半部，其尾型不同于土佐金，而与龙睛蝶尾类似，是通过与日本金鱼土佐金杂交选育出的一个品种。文蝶的观赏重点在尾鳍，要求尾鳍或大而舒展，或如翩翩舞裙；身形饱满，矫健有力，能平衡大尾鳍的阻力；眼珠齐平于眼眶，不能外凸。文蝶与宽尾琉金有相似之处，但宽尾琉金体高有变异，而文蝶基本属于正常金鱼的体形。文蝶目前的花色品种不多，主要是红色文蝶类和五花文蝶类。文蝶体质较强，易于饲养。由于对眼睛对称性要求并不严苛，其正品率相对比龙睛蝶尾要高一些，将来会成为龙睛蝶尾强有力的竞争对手。

Butterfly Tail Goldfish is a new goldfish variety appearing in recent years. Its head and body is like the fore body of Fantail Goldfish, but its tail is different from Tosakin Goldfish yet similar to Butterfly Moor. Butterfly Tail Goldfish.It is a variety through hybridization with the Japanese goldfish Tosakin Goldfish. The appreciation for Butterfly Tail Goldfish is its caudal fin, which should be large and extended or graceful like skirt. It is plump in body, strong and powerful, so it can balance the resistance from large caudal fin. The eyeballs should be horizontal with the eye sockets, but should not be convex. Butterfly Tail Goldfish has something in common with Ryukin Broad Tail, but Ryukin Broad Tail has a high hump with variation while Butterfly Tail Goldfish has the body shape of normal goldfish. At present, there are not many colors of Butterfly Tail Goldfish, mainly Red Fantail Goldfish Butterfly Tail and Calico Fantail Goldfish Butterfly Tail. Butterfly Tail Goldfish enjoys a good physique, so it is easy to feed. Because there is not a strict requirement for symmetry of eyes and eyeballs, the rate of graded goldfish is higher than that of Butterfly Moor Goldfish, and it will become a powerful competitor of Butterfly Moor Goldfish.

红黑文蝶
Red Black Fantail
Goldfish Butterfly Tail

花色类别 / Color •

红黑 / Red Black

品种类别 / Species •

文形 > 尾鳍变异类 > 文蝶型
Fantail Goldfish > Tail Fin Variation > Butterfly Tail

品种详述 / Species Description •

　　红黑文蝶是向红文蝶转色的过渡阶段的一个品种。文蝶和土佐金有血缘关系，因此，其尾鳍的变化也与土佐金有点接近，亲骨平直，末端呈现螺旋状翻转。如图例所示，其尾鳍亲骨呈现出较明显的螺旋状翻转，使得尾鳍如同盛放的牡丹花一般。由于尾鳍的缘故，文蝶游动较困难，因而它是一种对静态美的追求。

Red Black Fantail Goldfish Butterfly Tail is a transitional variety when it transits to the Red Fantail Goldfish Butterfly Tail. Because Butterfly Tail Goldfish has a blood relationship with Tosakin, the change of its caudal fin is a little similar to Tosakin, with flat and straight tail bone and spirally reversible end. As the legend shows, the tail bone of caudal fin obviously presents a spiral reverse, like a blooming peony. Due to its caudal fin, Butterfly Tail Goldfish swims difficultly, so it presents a pursuit of static beauty.

花文蝶

Calico Fantail Goldfish
Butterfly Tail

- **花色类别** / Color
  软鳞五花 / Calico

- **品种类别** / Species
  文形 > 尾鳍变异类 > 文蝶型
  Fantail Goldfish > Tail Fin Variation > Butterfly Tail

- **品种详述** / Species Description

　　花文蝶是从五花文蝶中选育而出的一个花色品种。图例所示的花文蝶，尾鳍亲骨的翻转程度不明显，但平展度较好，这一点和龙睛蝶尾又较相近。花文蝶体质较强，易于饲养，其多作为俯视欣赏，但侧视欣赏也有其看点，因此，自从培育出来以后，市场的接受度一直较高。

Calico Fantail Goldfish Butterfly Tail is a variety selected and bred from Calico Fantail Goldfish Butterfly Tail (Bluish Base). The Calico Fantail Goldfish Butterfly Tail shown in the figure has an in-apparent reverse of tail bone of caudal fin, but the caudal fin is open and flat enough. Calico Fantail Goldfish Butterfly Tail owns a good physique, so it is easy to feed. It is mainly appreciated from above, and sometimes from sides. Therefore, it has always been well accepted since it was bred.

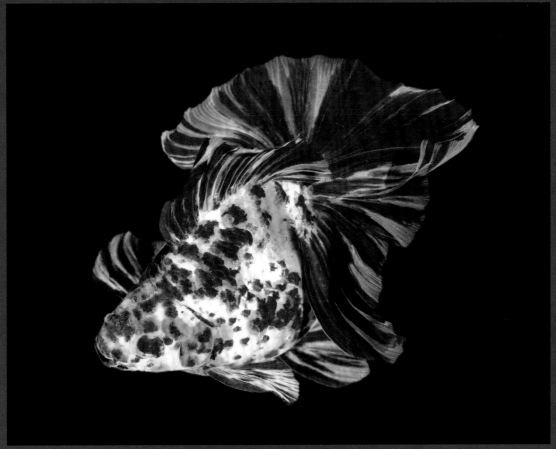

文形
Fantail Goldfish

∨

尾鳍变异类
Tail Fin Variation

∨

地金型
Jikin

地金原本是日本四大地方金鱼之一，1958年被尊为爱知县的天然纪念物，受到当地政府专门的保护和扶持，在日本享有崇高的地位。在日本还有地金的保育会，每年定期进行品评和展示。地金体形和原始的金鲫很接近，呈长梭形，又称柳叶身。但其尾鳍尾皿部向上翻转，使四片尾叶呈现"X"状形态，并与体轴呈90度垂直，如孔雀开屏，因此又称孔雀尾。俯视时，尾叶打开呈"U"字形的比呈"V"字形的要标准。地金中，以六鳞为贵。所谓六鳞，是类似中国的十二红色，再加上两抹鳃红。六鳞地金纯自然出现的概率非常之小，大多是通过刮鳞调色的手段得以实现，因而，与出云南京一样，六鳞地金也是通过做鱼体现创造美丽的乐趣。因为中日间文化的差异，此类做鱼的手法国人认同的不多。地金在国内繁殖培育已经许多代，其发展已有自己独特的一面，尤其是花地金，水准已经相当之高。

Jikin was originally one of the four local goldfishes of Japan. In 1958, it was revered as a natural monument of Aichi Prefecture. It is specially protected and supported by the local government and enjoys a high grade in Japan. Besides, there is Jikin Conservation Society in Japan, which regularly evaluates and displays Jikin Goldfish every year. In terms of body shape, Jikin is close to the original Golden Crucian Carp, like a spindle, and the body shape is also called willow leaf body. But, the tail pan of its caudal fin reverses upward, making the four caudal lobes present "X"; and it is 90 perpendicular to the body axle, like a peacock flaunting its tail, so it is also called peacock tail. When overlooked, the caudal fin presenting a "U" after spreading are more standard than those presenting a "V". Among Jikin, the six-scale ones are the most precious. The so-called six-scale is similar to Chinese twelve-red, coupled with two red gills. The purely natural appearance probability of six-scale Jikin is very small, and most of them are obtained through scaling and color matching. Therefore, like Cloudy Nankin, six-scale Jikin also reflects the fun of creating beauty through doing fish. Due to the difference between Chinese culture and Japanese culture, the means of doing fish is different from those of Chinese people. Jikin has been reproduced for many generations in China, and its development is unique; especially Calico Jikin enjoys a very high standard.

红白地金
Red White | Jikin

花色类别 / Color •
红白 / Red White

品种类别 / Species •
文形 > 尾鳍变异类 > 地金型
Fantail Goldfish > Tail Fin Variation > Jikin

品种详述 / Species Description •

　　图例所示为红白地金。它的操作手法是在地金鱼苗期，由青色逐渐向红色转色之前，刮去体表的鳞片，将其放在遮光的环境中，并在水中加入适当的药物，待其重新长出鳞片时，则会有部分金鱼呈现出我们所需要的颜色。地金的六鳞色作色手法，体现了日本民族为了追求极致而可以穷极一切手段的民族文化特性，不应用我们的文化视角去轻易褒贬。

The figures show Red White Jikin, whose operating method is as follows: at the frying period of Jikin, before the blue gradually turns to red, scrape the scales on the body, put it in the light-proof environment, and add appropriate medicine in the water; after the scales re-grow, some goldfish will have the color we need. The six-scale making method of Jikin reflects the national cultural characteristic of the Japanese nation pursuing the extreme by all means; and we should not easily appraise it from our own cultural prospective.

花色类别 / Color •

软鳞五花 / Calico

品种类别 / Species •

文形 > 尾鳍变异类 > 地金型
Fantail Goldfish > Tail Fin Variation > Jikin

花地金 | Calico Jikin

品种详述 / Species Description •

　　图例所示为花地金。地金在日本是一个小众品种，而花地金又是地金中的小众，过去一直鲜见五花色的地金，甚至于日本的许多资料中，对它都鲜有记载。近些年来，福州鱼场从日本引进花地金并加以培育，其品质大大得到提升，2014年福州金鱼展上花地金惊艳登场，得到包括日本专家在内的圈内人士的一致好评。

The figure shows Calico Jikin. Jikin is a minor variety in Japan, and Calico Jikin is a minor variety of Jikin. In the past, Calico Jikin was very rare, and it is hard to find from related documents in Japan. In recent years, Fuzhou Fish Farm has introduced Calico Jikin from Japan and bred it, so its quality has been greatly improved. At Fuzhou Goldfish Exhibition in 2014, Calico Jikin made an attractive fresh appearance, winning the praise of all the insiders, including the Japanese experts.

文形
Fantail Goldfish

∨

复合变异类
Compound Variation

∨

高身龙睛型
Telescope with High Hump

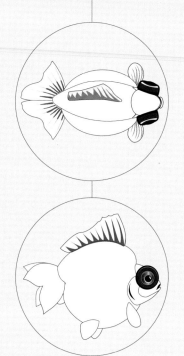

　　高身龙睛又称短尾龙睛，也有地方称之为出目金。出目金是日本对所有眼球凸出眼眶的龙睛的总称，所以，为了避免和龙睛混淆，我们采用了高身龙睛的名称。它是包含眼型变异类和背峰变异类的复合变异类品种。高身龙睛体形上与短尾琉金完全相同，只是两个眼球如龙睛凸出于眼眶之外。高身龙睛花色众多，体质强健，较易饲养，刚一推出就引发关注，也颇受市场青睐。高身龙睛适合南方鱼场的大规模坑塘化养殖，可与琉金等混养。

Telescope with High Hump is also called Telescope Short Tail, and also called Demekin in some places. Demekin is a generic term for all the Telescope Goldfish with eyeballs protruding out, so in order to avoid the confusion with Telescope Goldfish, we adopts the name of Telescope with High Hump. It contains the compound variation of eye variation and hump variation. In terms of body shape, Telescope with High Hump is totally the same as Ryukin Short Tail, but the two eyeballs of Telescope with High Hump protrude the eye sockets like Telescope. Telescope with High Hump have many colors, enjoy a good physique, so they are easy to feed. They attract much attention the moment they were brought out. It is suitable to feed Telescope with High Hump in the large-scale ponds in southern fish farms, and Telescope with High Hump can be fed together with Ryukin.

花色类别 / Color •

黑 / Black

品种类别 / Species •

文形 > 复合变异类 > 高身龙睛型

Fantail Goldfish > Compound Variation > Telescope with High Hump

品种详述 / Species Description •

黑高身龙睛 —— Black Telescope with High Hump

图例所示为黑高身龙睛的亚成体，虽然还没有发育成熟，但不失为一尾较优秀的金鱼。其外形浑圆，背峰和背鳍高耸，背鳍的第一鳍梗完整无残缺，重心平稳，无倾头或侧翻的倾向；腹部的黑色沉积还有所欠缺，这也是国鱼中许多黑色金鱼的普遍问题，可以通过饲养手段的改进，使其腹部颜色可以和背部一样厚如积墨。

The figure shows a sub-adult of Black Telescope with High Hump. Although it is not mature, it is already very outstanding. It has round body, high back hump and dorsal fin, intact first fin stem of the dorsal fin, steady center of gravity and has no tendency of head tilt or side turn. The black deposit on the belly is a little defective, which is common among many black goldfish in China and can be improved through feeding means. The color of the belly can be as dark as that on the back through improvement.

红白高身龙睛

Red White Telescope with High Hump

- **花色类别** / Color
  红白 / Red White

- **品种类别** / Species
  文形 > 复合变异类 > 高身龙睛型
  Fantail Goldfish > Compound Variation > Telescope with High Hump

- **品种详述** / Species Description

图例所示为红白高身龙睛。其红白花色很接近十二红，已非常难得；白色鳞片银白光洁，红色部分浓艳饱和，二者冷暖对比明显，互相映衬；尖头大眼很像漫画中的青蛙王子，呆萌可爱。红白高身龙睛是高身龙睛中最受欢迎的花色品种，可以和其他花色一起混养，也可以单独饲养一群，都有较好的观赏效果。

As is showing in the figure, it is a Red White Telescope with High Hump, its color is quite rare, close to "Twelve Red", with scales in white and red, complementary and contrasted with each other; the pointed head and eyes look like the frog in the comic books, so lovely and cute. The Red White Telescope with High Hump is most popular variety in all high-humped varieties. It not only can be raised alone, but also can be raised with other varieties, both are in high appreciation effect.

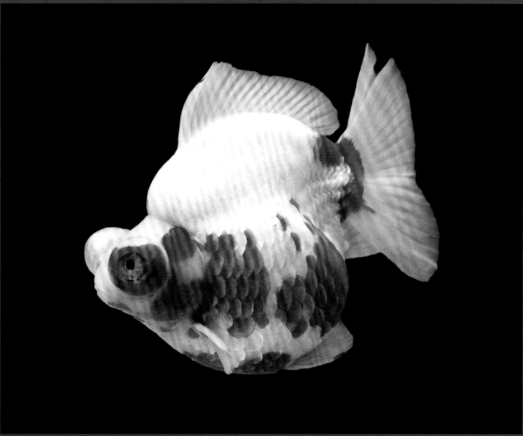

红黑高身龙睛

Red Black Telescope
with High Hump

- **花色类别** / Color
  红黑 / Red Black

- **品种类别** / Species
  文形 > 复合变异类 > 高身龙睛型
  Fantail Goldfish > Compound Variation > Telescope with High Hump

- **品种详述** / Species Description

　　图例所示为红黑高身龙睛。红黑高身龙睛在外部形态上极像短尾琉金，其背峰高耸，体态浑圆，尾鳍短小，腹部肥大以平衡重心。图例1所示红黑高身龙睛色彩极好，背部浓黑，腹部呈现橙红色，如夏日黄昏后的赤霞黑云一般。高身龙睛适合水族箱中侧视观赏，可以与红白短尾琉金或红白高身龙睛混合饲养，达到较好的视觉观赏效果。

The figures show Red Black Telescope with High Hump, it looks very much like Ryukin Short Tail, both of which have high back hump, round body, short caudal fin, and fat tommy to keep the gravity. The one in figure 1, its color is fantastic, jet back on the back and orange on the belly, just like the black cloud in the summer evening. High-hump telescope is better to be raised on the water tank together with Red White Ryukin with short tailor Red White Telescope with High Hump to reach better ornamental effect.

- **花色类别** / Color
  三色 / Red Black White

- **品种类别** / Species
  文形 > 复合变异类 > 高身龙睛型
  Fantail Goldfish > Compound Variation > Telescope with High Hump

- **品种详述** / Species Description

　　图例所示为三色高身龙睛。与前面所示的高身龙睛有所不同，此尾鱼为三色龙睛蝶尾中的小尾个体。小尾的蝶尾如果用饲养琉金的手法去培育，很容易出现高身的现象。三色高身龙睛颜色不易稳定，会向红白转色，适当控制温度、饲料中添加螺旋藻等方法，可以延缓这个转色的过程。

The figure shows a Red Black White Telescope with High Hump. Different from the above two, it is very short with butterfly tail among the butterfly varieties. If bed in the same method as breeding Ryukin, it is very easy to grow to a high hump shape. The color of it is quite changeable, normally will turn to red white. The way to control the color change is to manage temperature or add spirulina in forage.

花色类别 / Color •

软鳞红白 / Red White Matt

品种类别 / Species •

文形 > 复合变异类 > 高身龙睛型

Fantail Goldfish > Compound Variation > Telescope with High Hump

品种详述 / Species Description •

樱花高身龙睛
—Sakura Telescope with
High Hump (Red White Matt)

　　图例所示为樱花高身龙睛亚成体。樱花高身龙睛是软鳞花色，由五花高身龙睛培育而出。图例所示樱花高身龙睛体形优良，颜色方面其绯色面积过大，若能是白胜则会较完美；身上有硬质反光鳞点缀，适合水族箱侧视观赏，若能辅以适当灯光，则观赏效果更佳。

The figures show Sakura Telescope with High Hump (Red White Matt), it has calico Soft-Scale and is cultivated from Calico Telescope with High Hump. The one here has a good body shape, and it will be perfect if the part of scarlet on the body can be smaller. With the Hard-Scale reflecting, when put in the water tank and affiliated with proper lighting, it will provide a marvelous appreciation view.

231

花色类别 / Color •

软鳞五花 / Calico

品种类别 / Species •

文形 > 复合变异类 > 高身龙睛型

Fantail Goldfish > Compound Variation > Telescope with High Hump

品种详述 / Species Description •

花高身龙睛

Calico Telescope
with High Hump

　　图例所示为花高身龙睛亚成体，是由五花高身龙睛选育出的一个花色类型。花高身龙睛是非常适合水族箱养殖，侧视欣赏的一个品种。其外形圆润，体态憨拙。软鳞的花色在灯光的映射下会有大理石般的润泽感和肌理感。花高身龙睛适合群养观赏，也可以和短尾琉金混合搭配饲养，都有较好的视觉效果。

The figure shows a sub-adult of Calico Telescope with High Hump, which is a variety selected and bred from Calico Telescope with High Hump (Bluish Base). Calico Telescope with High Hump is very suitable for aquarium feeding and appreciation from sides. It boasts of round shape and naive posture. In the light, the color of Soft-Scale will have a marble-like moist and textural feeling. Calico Telescope with High Hump is suitable for group feeding and appreciation, and can also be fed together with Ryukin Short Tail. Both of them will have a good visual effect.

重彩高身龙睛

Telescope with High Hump
(with Intense Colors)

- **花色类别** / Color
  软鳞五花 / Intense Colors

- **品种类别** / Species
  文形 > 复合变异类 > 高身龙睛型
  Fantail Goldfish > Compound Variation > Telescope with High Hump

- **品种详述** / Species Description

　　图例所示为重彩高身龙睛，是从荧鳞高身龙睛中选育出的一个花色类型。其背部白色软鳞微微带有一些红晕，如寿山的芙蓉石，又如薛涛的"桃花笺"，下面泼洒墨彩，任由其自然地晕染渗化；体形上由于是亚成体，还未完全发育，但可以看出后续有较好的发展空间。

The figure shows Telescope with High Hump (with Intense Colors) which is a variety cultivated from Fluorescent Scale Telescope with High Hump. Its white Soft-Scale at back has certain red halo, like Ross quartz in Shou Mountain, or Pink Paper made by Xue Tao, a Chinese traditional painting. Though it's a sub-adult which is not fully developed, we can perceive that it has great potential in the future.

文形
Fantail Goldfish

∨

复合变异类
Compound Variation

∨

高头球型
Pompon Oranda

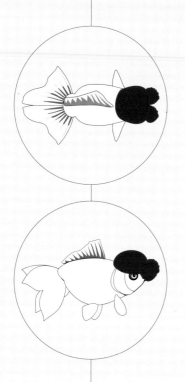

高头球是头型变异类和鼻膜变异类叠加的复合变异类品种。高头球大约在20世纪50年代被选育出，但一直属于小众品种，没有大规模发展起来。高头球又称帽子球，多为文形金鱼中虎头型与文球杂交而得。鼻瓣膜变异类的绒球，在遗传学上属于显性遗传，所以，许多复合变异类是以绒球搭配其他变异形式而存在的。高头球的球由于发育受到顶茸的限制，因此其发育度不及单纯的绒球型变异，但球体相对紧实。高头球体质强健，较易饲养，其花色品种众多，除了常见的红高头球、红白高头球、五花高头球外，还培育出蓝高头球、紫高头球、雪青高头球、黑白高头球和三色高头球等新花色类型。高头球目前主要产地为福州、上海和扬州等地。

The Pompon Oranda is the compound variety of head variation and nasal variation. Although it has been developed in the 50s of twentieth century, it is only favored in small group. Named also Cap Head, with Fantail Goldfish, it is crossed mostly by Tigerhead and Pompons. The nose pompon mutated from the nasal valve is autosomal dominant. According to this, many other composite variations with pompon are also the hybrid of the Pompon Oranda with other varieties. Although the pompom is restricted in growth than the others with pure pompon type, but the head pompon is relatively more compact than the latter one. The Pompon Oranda is easy to breed and has many varieties, except the Red Pompon Oranda, Red White Pompon Oranda, Calico Pompon Oranda, the Oranda of Blue Pompon, Chocolate Pompon, Lilac Pompon, Black White Pompon as well as the Tri-Colored Pompon are also developed. The origin of Pompon Oranda is mainly in Fuzhou, Shanghai and Yangzhou.

花色类别 / Color •

红 / Red

品种类别 / Species •

文形 > 复合变异类 > 高头球型

Fantail Goldfish > Compound Variation > Pompon Oranda

品种详述 / Species Description •

红高头球 —— Red Pompon Oranda

　　红高头球是一个古老的品种，大约产生于20世纪50年代，是由文形金鱼中红虎头与文球杂交培育而出的。红高头球头部顶茸平整而发达，顶茸上有阴刻的类似王字样花纹；嘴上部有两个绒球，高头球的绒球发育虽不及文球等发达，但紧实圆润；体质强健，较易饲养。红高头球俯视侧视观赏均可。

Red Pompon Oranda is a very ancient breed of goldfish, generated on 1950s of twentieth century, hybridized by Tigerhead and Fantail Goldfish with Pompons. The pompons on the head are well organized and developed, with a shape bearing the Chinese characteristic "Wang". Above the snout, there are two hoods, not as developed as Fantail Goldfish with Pompons, but more tight and round. It is both ok to side view or down view the Red Pompon Oranda. With the strong physique, it is very easy to be bred. Red Pompon Oranda is appropriate to be viewed both from top and sides.

花色类别／Color •

红白／Red White

品种类别／Species •

文形 ＞ 复合变异类 ＞ 高头球型

Fantail Goldfish ＞ Compound Variation ＞ Pompon Oranda

品种详述／Species Description •

红白高头球是高头球中红白花色类型，红白高头球颜色变化丰富，是高头球中最受欢迎的花色类型之一。图例1、4所示为赤胜花色，其顶茸及绒球发育良好，背鳍高耸挺拔，尾鳍宽大飘逸，体态优美。图例2花色为白胜花色，白身红尾，若两绒球为红色则更加完美。图例3为红顶花色，朱砂眼配以红白双色的鸳鸯球，是可遇而不可求的佳色妙品。

Red White Pompon Oranda is red white among Pompon Oranda, with its changeable color on Pompons, is one of the most favorite colors. In figure 1, 4, they are Akagachisarasa Oranda with well-developed pompon with high dorsal fin, wide caudal fin, so elegant and beautiful. Figure 2 is Shirogachisarasa Oranda with white body and red tail. If the two hoods are also red, it will be much fancier. In figure 3, the Oranda is red caped, with the eye balls in cinnabar and two pompons respectfully in red and white, The eyes like this is called "Yuanyang" eyes. (Yuanyang, Mandarin ducks in Chinese, usually refers to a mixture of two different types of the same category). A Red White Pompon Oranda like this can be ranked as rare.

红白高头球
Red White｜Pompon Oranda

五花高头球

| Calico Pompon Oranda (Bluish Base)

- **花色类别** / Color

  软鳞五花 / Calico (Bluish Base)

- **品种类别** / Species

  文形 > 复合变异类 > 高头球型

  Fantail Goldfish > Compound Variation > Pompon Oranda

- **品种详述** / Species Description

  五花高头球是高头球中软鳞五花色类型，图例所示五花高头球头茸和绒球紧实，体态优美，背鳍高耸挺立，尾鳍舒展飘逸。其面部红色，并配以白球，十分漂亮。腹部大片橙红色是五花花色中的大忌，如果为白腹配以蓝背碎花会比较完美。五花高头球比较稀少，好的花色则更加珍稀。

  The Calico Pompon Oranda (Bluish Base) is Soft-Scaled type. Showing in figure below, it is with a compact and tight pompon, a high dorsal fin and a wide stretching caudal fin. Red cheeks, accompanied with white pompons is so beautiful. It is a taboo if the belly is red in large area. However, on the contrast, if white belly is with blue spots it can be regarded as perfect. Normally it is very rare that a Calico Pompon Oranda (Bluish Base), not to mention those with nice colors.

文形
Fantail Goldfish

∨

复合变异类
Compound Variation

∨

皇冠珍珠型
Crown Pearlscale

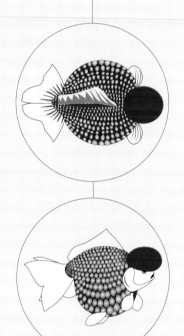

皇冠珍珠又称帽子珍珠鳞或鹅头珍珠，它是皮球珍珠与头型变异类金鱼通过杂交选育出的一个品种，是具有鳞片变异类和头型变异类特征的复合变异类品种。和皮球珍珠类似，皇冠珍珠身体也呈皮球状，其全身覆盖由石灰质沉积的珍珠状鳞片，头顶有发达顶茸，单瓣呈蛋形，双瓣呈鸡心形，如同头戴通天冠，故而称之为皇冠珍珠。皇冠珍珠的出现大约在20世纪50年代，起初未被重视，后传入日本，1978年正式将其定名为滨锦。滨锦由于引入的是早期的帽子珍珠，所以其顶茸呈颗粒状分布。而后期，福州鱼场对帽子珍珠加以改造，逐渐形成了现在皇冠珍珠的模样。皇冠珍珠后来逐步完善花色，形成了红色、紫色、蓝色、黑白、红白、三色、麒麟和五花等诸多花色。

Crown Pearlscale, also named Caped Pearlscale or Goosehead Pearlscale, is a hybrid of Golfball Pearlscale and Head Variations varieties, belonging to the Compound Variation of scale and head. Like the Golfball Pearlscale, the Crown Pearlscale has a round body as well, covered with calcareous sediments pearlscales. This breed has very developed pompon, either in Egg-Fish or in heart shape (seen in two-pompon ones). With the pompon looking like a crown, it is so called Crown Pearlscale. Initially, it has not caught people's eyes very much when it was developed in 1950s. After introduced to Japan in 1978, it was renamed as Hamanishiki. Hamanishiki belongs to the Caped Pearlscal of early stage, so the pompon is granule interspersed. Later, gradually, in Fuzhou area, Caped Pearlscale was improved and developed into current looking as it is right now. There are many other colors cultivated, such as red, chocolate, blue, black white, red white, tri-color, Kirin and calico.

花色类别 ／ Color ●

青 ／ Green

品种类别 ／ Species ●

文形 ＞ 复合变异类 ＞ 皇冠珍珠型
Fantail Goldfish ＞ Compound Variation ＞ Crown Pearlscale

品种详述 ／ Species Description ●

青皇冠珍珠 ／ Green Crown Pearlscale

　　图例所示为青皇冠珍珠，多从五花皇冠珍珠中出现，红皇冠珍珠中也会出现。青皇冠珍珠在越冬时，会逐渐转色为黑皇冠珍珠，也有部分会转为红皇冠珍珠，也有部分个体终生为青色个体。图例所示青皇冠珍珠的珍珠鳞饱满，排列整齐。独瓣顶冠非常发达，很吸引人眼球。青皇冠珍珠在转向黑色时，光照度要好，这样的黑色鱼色彩才能越发浓郁。

Green Crown Pearlscale, as is showing in figures, is cultivated from Calico Crown Pearlscale (Bluish Base) or Red Crown Pearlscale. During winter, it mostly turns black, sometimes turns red, some turns green.The one here is with full scale and good looking pompon, so attractive. When a Green Crown Pearlscale turns to black, the more lighting there is, the darker it will turn to.

黑皇冠珍珠
Black Crown Pearlscale

- **花色类别** / Color

  黑 / Black

- **品种类别** / Species

  文形 ＞ 复合变异类 ＞ 皇冠珍珠型

  Fantail Goldfish ＞ Compound Variation ＞ Crown Pearlscale

- **品种详述** / Species Description

  黑皇冠珍珠是广受大家喜爱的一个皇冠珍珠花色品种。黑皇冠珍珠是从红皇冠珍珠或五花皇冠珍珠中的青色个体中培育出的，前者黑色不易稳定，会逐渐向红色转变，而后者会由青色逐渐转为黑色，且黑色愈来愈浓烈。黑皇冠珍珠要求体形如球，珍珠鳞颗粒饱满，排列整齐，尾柄处尤其不能缺鳞。头冠发达，以双瓣鸡心形为最佳。

  Black Crown Pearlscale is one of the most popular breeds among all Crown Pearlscale. It is cultivated from Red Crown Pearlscale and Calico Crown Pearlscale (Bluish Base). The first one is easy to change from black to red, and the latter one is likely to change from green to black, and will become more and more dark as it's growing. It is the best if the body is as round as a ball, scales compact and organized, complete caudal fin, and the pompon is well developed with twin hearts shape.

红白皇冠珍珠

Red White
Crown Pearlscale

- **花色类别** / Color

  红白 / Red White

- **品种类别** / Species

  文形 ＞ 复合变异类 ＞ 皇冠珍珠型

  Fantail Goldfish ＞ Compound Variation ＞ Crown Pearlscale

- **品种详述** / Species Description

  红白皇冠珍珠在皇冠珍珠中以花色著称，最为名贵的是类似鹤顶红全身洁白如玉，唯有顶冠赤红，且呈鸡心形的红顶皇冠珍珠。红顶皇冠珍珠没有稳定遗传，往往可遇不可求。图例所示红白皇冠珍珠为二段红白色，以背鳍为界将前后分为两段红色。其珍珠鳞特别饱满，颜色艳丽，是一尾十分优秀的皇冠珍珠。

  Red White Crown Pearlscale is famous for its color. The most valuable one is heart shaped, much like Red Cap Oranda whose whole body is as pure as white jade, except for red in the hood. Since the crown shaped red pompon is not stably genetic, it is very rare to see many of this breeding. The one in the illustration is very valuable, with two-level red white, red in front and back divided by the dorsal fin, while the scales are full and dazzling.

红黑皇冠珍珠

Red Black
Crown Pearlscale

- **花色类别** / Color

  红黑 / Red Black

- **品种类别** / Species

  文形 > 复合变异类 > 皇冠珍珠型

  Fantail Goldfish > Compound Variation > Crown Pearlscale

- **品种详述** / Species Description

  图例所示为红黑皇冠珍珠，是黑皇冠珍珠向红皇冠珍珠转色的过渡阶段品种。图例所示红黑皇冠珍珠体形丰满，背鳍高耸挺立，尾鳍舒展，珍珠鳞颗粒饱满、全身以侧线分界，上半部为黑色，下半部为橙红色，眼口均为黑色，显得俏皮可爱。冠顶为单瓣类型，比较发达。

  The figure shows a Red Black Crown Pearlscale, which is the transition between Black Crown Pearlscale and Red Crown Pearlscale. The one in the picture is with a fat body, high dorsal fin and stretching caudal fin, compact and tight pearlscale. Separating by the lateral line of its body, the upper part is black and the lower part is red orange with both snout and eyes in black, which make it look so adorable. The pompon of it is single type, comparatively developed.

三色皇冠珍珠

Red Black White
Crown Pearlscale

- **花色类别** / Color
  三色 / Red Black White

- **品种类别** / Species
  文形 > 复合变异类 > 皇冠珍珠型
  Fantail Goldfish > Compound Variation > Crown Pearlscale

- **品种详述** / Species Description

  图例所示为三色皇冠珍珠，是从五花皇冠珍珠中培育出的一个花色品种。三色皇冠珍珠的顶冠较红白皇冠珍珠的要小，且多为双瓣，比较接近早期帽子珍珠的模样。另一种三色皇冠珍珠是从红白皇冠珍珠中选育而出的，颜色较浓郁，但褪色比例较高，饲养较少。

  The figure shows a Red Black White Crown Pearlscale, cultivated form the Calico Crown Pearscale (Bluish Base); it has smaller and twin caps, looking like the Caped Pearscale during initial stage. Besides this, another variety is selected and bred from Red White Pearlscale, with strong color, but easy to fade, therefore it is not so common to be kept by raisers.

花色类别 / Color •

珍珠鳞五花 / Kirin

品种类别 / Species •

文形 > 复合变异类 > 皇冠珍珠型

Fantail Goldfish > Compound Variation > Crown Pearlscale

品种详述 / Species Description •

麒麟皇冠珍珠
Kirin Crown Pearlscale

　　图例所示为麒麟皇冠珍珠，是由五花皇冠珍珠中选育而出的。其顶冠非常发达，多双瓣型，红顶的称为红顶墨麒麟，白冠的称玉顶墨麒麟，十分抢眼。麒麟皇冠珍珠幼鱼期为五花色，三四月龄后，身上墨色开始逐渐加深，面积扩大，直至铺满全身。

The figure shows a Kirin Crown Pearlscale, selected and bred from Calico Crown Pearlscale (Bluish Base), normally with good head pompon in twin hearts shape. The red head pompon ones are called Black Kirin with Red Pompon, and the white ones are called Black Kirin with Jade Pompon, both are quite outstanding. The fries of Kirin Crown Pearscale are in Calico color, then after three or four Months' later, the black color will gradually extend to the whole body.

花色类别／Color •
珍珠鳞五花／Calico

品种类别／Species •
文形 > 复合变异类 > 皇冠珍珠型
Fantail Goldfish > Compound Variation > Crown Pearlscale

品种详述／Species Description •

　　图例所示为花皇冠珍珠。花皇冠珍珠是由五花皇冠珍珠中选育而出。图例所示金鱼顶冠异常发达，单瓣型，橙色的顶冠上还点缀墨色碎点花，十分漂亮。其体形圆润饱满，珍珠鳞排列整齐。遗憾的是色彩不够爽朗，有些灰暗。

As is shown in the figure, Calico Crown Pearlscale is also selected and bred out of Calico Crown Pearlscale (Bluish Base). Its cap is extremely developed, single piece, orange with black spot as ornament. In addition, the body is round with organized pearlscale. The disadvantage is that its color is not bright enough.

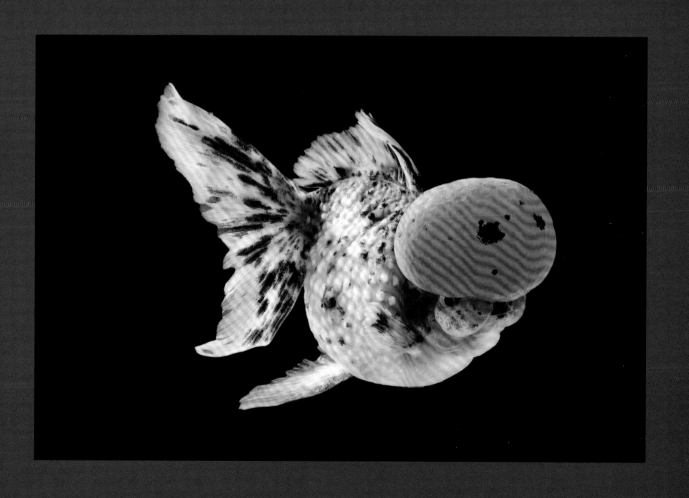

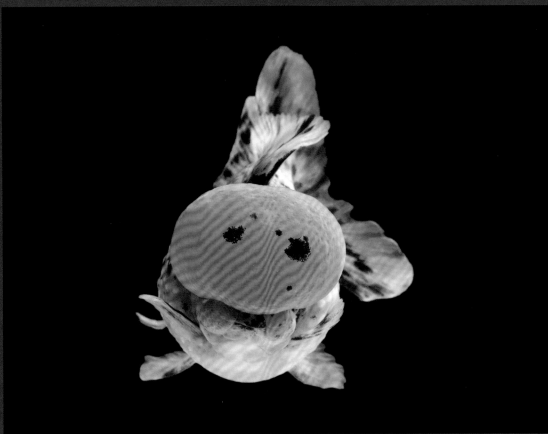

五花皇冠珍珠

—— Calico Crown Pearlscale (Bluish Base)

花色类别 / Color •

珍珠鳞五花 / Calico (Bluish Base)

品种类别 / Species •

文形 ＞ 复合变异类 ＞ 皇冠珍珠型
Fantail Goldfish ＞ Compound Variation ＞ Crown Pearlscale

品种详述 / Species Description •

　　图例所示为五花皇冠珍珠。五花皇冠珍珠是皇冠珍珠中体色较难培育的一个花色品种，其蓝色部分较易丢失。五花皇冠珍珠顶冠较其他花色要小些，多双瓣头型。五花皇冠珍珠主要产自福州、南京、苏州等地。饲养时要保持水质清新，最好用硬水，这样有利于珍珠鳞中钙和磷的沉积。珍珠鳞在操作时要谨慎，一旦脱落，恢复期会比较长。

The figures show Calico Crown Pearlscale (Bluish Base). It is a rare breed cultivated from Crown Pearlscale. The blue color is very easy to disappear; it has smaller head pompon than other varieties of Crown Pearlscale, mostly are with twin hearts shape. The origins are mostly based in Fuzhou, Nanjing and Suzhou. The water for breeding must be hard water and as clean as possible, which is good for the sediment of calcium and phosphorus of the scales. During operation, people must very prudent, since the scale, once falls off, will take long to get recovered.

文形
Fantail Goldfish

∨

复合变异类
Compound Variation

∨

龙睛蝶尾型
Butterfly Moor

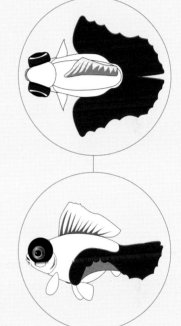

　　龙睛蝶尾俗称蝶尾，是中国金鱼优秀代表品种之一。龙睛蝶尾是从龙睛中定向选育而出，是眼型变异类与尾鳍变异类叠加的复合变异类品种。龙睛蝶尾起初为小尾，后来经过不断筛选而逐渐发展成今天这种形态，但在蝶尾繁育时，还是会出现许多小尾的个体。龙睛蝶尾尾型的变化有人误以为和日本的土佐金类似，其实二者有本质的区别。土佐金的亲骨平展，呈横向的螺旋状翻转，而龙睛蝶尾多为向眼睛处反翘。龙睛蝶尾主要有3种类型：以江苏如皋为代表培育的反翘型、以福建福州为代表培育的高身平蝶型和以甘肃天水为代表培育的剪刀尾型，但前二者接受度高，而天水蝶认同度较低。由于以前如皋地区偏重淀粉型饲料，如皋蝶尾虽身形饱满，但眼球较小，而如皋独特的饲育手法，成就了如皋蝶尾尾先可以触及眼球甚至过嘴的超级发达状态。福州蝶尾饲养偏重动物型蛋白饲料，加上水深较深，造就福州蝶尾高身大眼平尾展的特点。福州蝶尾雍容华贵，炫丽多彩，有水中牡丹的雅称。有诗云："清风拂来疑花影，水痕淡去弄双蝶。"这是对龙睛蝶尾最真切的描绘。

Butterfly Moor also called Butterfly Tail, is one of the leading species of Chinese goldfish. It is selected from Dragon Eyes, and is the composition of the eye variations and dorsal fin variations. This variety starting with small tail, has been gradually developed to the current big shape, although, during the reproduction, the small tail type still appears. The variation of it are mistakenly considered the same as Tosakin in Japan. However, they are actually completely different. The tail bone of Tosakin is in heliciform while for Butterfly Moors, it turns out towards the eyes. There are three main types which are exemplified by trapezium tail in Rugao of Jiangsu, High Hump Flat Butterfly Moor in Fuzhou of Fujian, and the Fork Tail Butterfly Moor in Tianshui of Gansu. Comparatively speaking, the previous two are better recognized than the third one. With much starch in the forage, Rugao Butterfly Moor is with round body but small eyes. Attribute to the special breeding, the tail of Rugao Butterfly Moor is capable to touch the eyes or even extend over the snout. As for Fujian Butterfly Moor, the forage is more with animal protein, and the water for breeding is deep, based on which it is characterized by high hump, big eyes and flat tail. Fuzhou Butterfly Moor is so elegant and dazzling that it is reputed as peony in water. "The wind blows to suspected flowers, water marks to get double light" as is said in the poem.

红龙睛蝶尾 | Red Butterfly Moor.

- **花色类别** / Color

  红 / Red

- **品种类别** / Species

  文形 > 复合变异类 > 龙睛蝶尾型

  Fantail Goldfish > Compound Variation > Butterfly Moor

- **品种详述** / Species Description

　　图例所示为红龙睛蝶尾。红龙睛蝶尾是龙睛蝶尾中最普遍的一个花色品种。如皋出产的红龙睛蝶尾，尾先可以触及双目，甚至可以过嘴，真是令人叹为观止。图例中的红龙睛蝶尾身形饱满，比例适中，全身火红，如丹霞彤云，尽显雍容华贵之姿。红龙睛蝶尾较易饲养，常用大型白陶瓷盆浅水玩赏。

The figures show the Red Butterfly Moor, one of the most common butterfly varieties. The breed produced in Rugao, is amazing because the tail can reach to the eyes or even exceeds the snout. The one in the picture with full proper portion body, as red as rosy cloud, presents a noble elegance. It is easy to raise, normally in a white pottery bowl.

黑龙睛蝶尾 | Black Butterfly Moor

- **花色类别** / Color
  黑 / Black

- **品种类别** / Species
  文形 > 复合变异类 > 龙睛蝶尾型
  Fantail Goldfish > Compound Variation > Butterfly Moor

- **品种详述** / Species Description

　　图例所示为黑龙睛蝶尾亚成体，全身如着黑绒缎礼服，又似水中盛开的黑牡丹。黑龙睛蝶尾又称墨蝶，是黑色系列金鱼中颜色较稳定的一个类型，且年龄越大，颜色越深。黑龙睛蝶尾培育时间较早，初多小尾，后来逐渐被大尾所取代。五花龙睛蝶尾苗中，会有相当比例青色的硬鳞苗个体转为黑龙睛蝶尾，十分有趣。黑龙睛蝶尾以轻巧灵动为贵，宜在豆青釉彩或天青釉彩的陶瓷浅盆内玩赏。

The figures show the Black Butterfly Moor. It looks like wearing a black tux, and also like the black blossom peony in water. Black Moor is another name, among other varieties of this serial, its color is relative stable, and will turn stronger during growing process. The cultivation began early, staring with small tail variety, and was gradually replaced by the big tail variety. It is very interesting that there is a high potential the fry of the Calico Butterfly Moors (Bluish Base) with Hard-Scale may transit into Black Butterfly Moors later. Black Butterfly Moor is agile and graceful, so it is better to keep them in pea green glaze or azure glaze ceramic basin.

蓝龙晴蝶尾 | Blue Butterfly Moor

- **花色类别** / Color
  蓝 / Blue

- **品种类别** / Species
  文形 > 复合变异类 > 龙晴蝶尾型
  Fantail Goldfish > Compound Variation > Butterfly Moor

- **品种详述** / Species Description

　　图例所示为蓝龙晴蝶尾。蓝龙晴蝶尾也是比较古老的一个花色品种，出现年代较早。蓝龙晴蝶尾和许多蓝色系列的金鱼一样，会在高温环境下出现褪色现象，变化出黑白色，但这种黑白色很难持久，和真正的熊猫蝶尾是不同的花色品种。图例所示蓝龙晴蝶尾身形过于臃肿，失去了翩翩蝴蝶的那份轻盈感和灵动感。

The figures show the Blue Butterfly Moor. It is a very ancient variety. Like other blue varieties, the color fades under high temperature and turns to black white. But it won't last long, and it should be distinguished from the Panda Butterfly Moors. The ones in the figures are so fat that they have lost the agility and elegance of a butterfly.

花色类别 / Color •

紫 / Chocolate

品种类别 / Species •

文形 ＞ 复合变异类 ＞ 龙睛蝶尾型
Fantail Goldfish ＞ Compound Variation ＞ Butterfly Moor

品种详述 / Species Description •

紫龙睛蝶尾
—— Chocolate Butterfly Moor

　　图例所示为紫龙睛蝶尾。紫龙睛蝶尾会根据饲养水质呈现出赤铜色或黄铜色的变化。过去的紫龙睛蝶尾身形普遍偏小，但背鳍高挺，尾鳍宽大，给人以灵秀之美。现在的紫龙睛蝶尾身形变大，体高也增加，有一定的视觉冲击力，但却失去了那份灵秀。因此，在对一个花色创新、品种塑形的时候，还是需要多加斟酌。

As is shown in the figure, it is Chocolate Butterfly Moor. With variant water quality, it shows a color with red copper or yellow copper. This variety used to be very small, usually with high hump and wide caudal fin. Presently, it becomes much bigger and higher, although it has provided with a strong contrast on vision, yet lost the initial agility. Therefore, it need more considerations when selecting or developing a new variety.

红白龙睛蝶尾 | Red White Butterfly Moor

- **花色类别** / Color
  红白 / Red White

- **品种类别** / Species
  文形 > 复合变异类 > 龙睛蝶尾型
  Fantail Goldfish > Compound Variation > Butterfly Moor

- **品种详述** / Species Description

　　图例所示为红白龙睛蝶尾。在大多数花色中，红白一定是最受青睐的。红白龙睛蝶尾中，最著名的就是十二红蝶尾，所谓十二红就是指两胸鳍、两腹鳍、两臀鳍、两尾鳍、双眼、一个背鳍加一个吻，一共十二处红色，其他地方洁白无瑕。十二红蝶尾最难的就是所有的鱼鳍要红到梢，不能有一点白丝，眼球要红到眼根，唇红更是不能缺失，正如《红楼梦》中形容冷香丸的制作，难得"可巧"二字。除此以外，全身洁白，唯有双眼赤红的玛瑙眼，又称双灯照雪，也是红白龙睛蝶尾中的极品。

The figures show Red White Butterfly Moor. In common sense, red white is the most popular to fanciers. Of these varieties, the most famous one is called Twelve-fin Red Butterfly Moor. Twelve-fin refers to two chest fins, two belly fins, two analfins , two caudal fins, a pair of eyes, a dorsal fin and a snout, totally twelve red, with other area in jade white. It is extremely rare that the fins should be red till the end, without a tiny white line; the eye is red till the orbit, and indispensably the snout should be red too. It is as rare and value as of the scent balls, recorded in the book *The Dream of Red Mansions*. In addition, the whole body is snow white except for the agate-red eyes, just like two red lanterns on the snow ground. It is regarded as the top grand on Red White Butterfly Moor.

红黑龙睛蝶尾

Red Black
Butterfly Moor

- **花色类别** / Color
  红黑 / Red Black

- **品种类别** / Species
  文形 > 复合变异类 > 龙睛蝶尾型
  Fantail Goldfish > Compound Variation > Butterfly Moor

- **品种详述** / Species Description

　　图例所示为红黑龙睛蝶尾。红黑龙睛蝶尾是黑龙睛蝶尾向红龙睛蝶尾转色的过渡阶段品种。图例中是一个红黑龙睛蝶尾的巧色，它在背鳍两边各有一对称的圆形红斑，如同两个大眼睛。可巧的是，自然界中，许多蝴蝶的翅膀上也有类似眼型的花斑，这种巧色花斑极难遇见，只是可惜目前还没有办法将其固定下来。

The figures show the Red Black Butterfly Moor, a transit variety from Black Butterfly Moor to Red Butterfly Moor. It is in ingenious red, with two round red spots besides the dorsal fin like two big eyes. Coincidently, in the natural world, these kinds of spot can also be seen in many butterflies. Despite of its rareness, there is no way to keep it stable at present.

黑白龙睛蝶尾

Black White
Butterfly Moor

- **花色类别** / Color
  黑白 / Black White

- **品种类别** / Species
  文形 > 复合变异类 > 龙睛蝶尾型
  Fantail Goldfish > Compound Variation > Butterfly Moor

- **品种详述** / Species Description

　　图例所示为黑白龙睛蝶尾，这种花色是从蓝龙睛蝶尾中选育而出的，和真正的黑白色熊猫龙睛蝶尾有所差异。图例所示为如皋出产的花色品种，反翘型尾特点突出，尾先可以及目，尾鳍边缘有如同日本土佐金朝颜般的尾齿花边，十分美丽。其身上花色的白色部分不够干净，光泽度欠佳。眼部若能镶嵌对称的银环，则会更加完美。黑白龙睛蝶尾可以通过降低水温，在饵料中适量添加螺旋藻等方法延长其褪色的过程。

As is showing in the figures, the Black White Butterfly Moor is selected variation from Blue Butterfly Moor, different from White Panda Butterfly Moor. The one in the figure produced in Rugao, has a very unique feature on tail. Its tail edge can cocked back to the eyes. The caudal fin has beautiful lace end like Japan Tosakin. However, the white is not pure and bright enough. It can be more attractive if two silver rings grow at the sides of the eyes. To stop the color fading, the water should be controlled in low temperature and spirulina should be added into the forage.

紫白龙睛蝶尾
Chocolate White Butterfly Moor

- **花色类别** / Color
  紫白 / Chocolate White

- **品种类别** / Species
  文形 > 复合变异类 > 龙睛蝶尾型
  Fantail Goldfish > Compound Variation > Butterfly Moor

- **品种详述** / Species Description

　　图例所示为紫白龙睛蝶尾。紫白龙睛蝶尾有的是从紫龙睛蝶尾中选育而出，也有的是从紫蓝龙睛蝶尾中选育出，而笔者也发现从熊猫龙睛蝶尾的子代中，也会分离出紫白龙睛蝶尾的个体。紫白龙睛蝶尾以福州和如皋两地培育的最具代表性，并曾多次在各类金鱼比赛中摘金夺银。紫白龙睛蝶尾由于褪色比例较高，因此饲养较少，属于珍稀花色品种。

Chocolate White Butterfly Moor, as is showing in the figures, is cultivated from Chocolate Butterfly Moor, and sometimes is also from Chocolate Blue Butterfly Moor. The writer also realized that the offspring of Panda Butterfly Moor can also develop into Chocolate White Butterfly Moor. The most representative varieties, produced in Fuzhou and Rugao, have won the gold or silver medals during many goldfish competitions. Due to the high chance of color fading, it is not so common to be raise, so is very rare.

花色类别 / Color •

紫蓝 / Chocolate Blue

品种类别 / Species •

文形 > 复合变异类 > 龙睛蝶尾型

Fantail Goldfish > Compound Variation > Butterfly Moor

品种详述 / Species Description •

紫蓝龙睛蝶尾

Chocolate Blue
Butterfly Moor

图例所示为紫蓝龙睛蝶尾。紫蓝龙睛蝶尾由蓝龙睛蝶尾选育而出，紫龙睛蝶尾中也会有个体褪出紫蓝花色。紫蓝龙睛蝶尾属于小众花色品种，饲养数量较少。图例所示紫蓝龙睛蝶尾与十二红有异曲同工之妙，其全身蓝色，唯独所有的鱼鳍、眼睛和吻都是紫色的，所以也有称之为十二紫的，同样也是可遇不可求的极品花色。

As is showing in the illustration, Chocolate Blue Butterfly Moor is the breed selected from Blue Butterfly Moor, a few of them also from Chocolate Butterfly Moor. Since it is only received by a minor group, it is not raised in large scale. The Chocolate Blue Butterfly Moor in the picture, as delicate as twelve red, except the blue body, are all in Chocolate Blue in the fins, eyes and the snout; So the name twelve chocolate is after that, as precious as the twelve red.

三色龙睛蝶尾 Red Black White Butterfly Moor

- **花色类别** / Color
  三色 / Red Black White

- **品种类别** / Species
  文形 > 复合变异类 > 龙睛蝶尾型
  Fantail Goldfish > Compound Variation > Butterfly Moor

- **品种详述** / Species Description

　　图例所示为三色龙睛蝶尾。三色龙睛蝶尾由黑白龙睛蝶尾选育而出，呈现黑、红、白3种颜色，但其黑色部分保持不易，会褪色成红白龙睛蝶尾。南京的金鱼业者曾经用黑白龙睛蝶尾分别与红白龙睛蝶尾以及紫龙睛蝶尾进行杂交，培育出两只眼睛为红色，而身体为黑白色的龙睛蝶尾，称为火眼熊猫，其色相对稳定一些，也属于三色龙睛蝶尾中的珍稀花色品种。

The figures show Red Black White Butterfly Moor, is selected and bred from Black White Butterfly Moor, with black, red and white color on the body; the black is hard to retain, so it might change to Red White Butterfly Moor. The goldfish raisers have hybridized Black White Butterfly Moor with Red White Butterfly Moor or with Chocolate Butterfly Moor, the Black White Butterfly Moor is cultivated, named Golden Eye Panda Butterfly Moor. The color of it is comparatively stable, and is regarded as a precious variety among Red Black White Butterfly Moor.

樱花龙睛蝶尾
Sakura Butterfly Moor
(Red White Matt)

- **花色类别** / Color
  软鳞红白 / Red White Matt

- **品种类别** / Species
  文形 > 复合变异类 > 龙睛蝶尾型
  Fantail Goldfish > Compound Variation > Butterfly Moor

- **品种详述** / Species Description

　　图例所示为樱花龙睛蝶尾，是软鳞红白花色，由五花龙睛蝶尾培育而出。软鳞花色的金鱼相较硬鳞花色，更有一种如同寿山芙蓉冻石般的润泽感。软鳞花色中的红称之为绯，白色称之为地，图例中的樱花龙睛蝶尾绯与地的比例恰到好处。绯应该和地有一种渗透感，这样感觉色彩是流动的、有生命力的。

The figures show the Sakura Butterfly Moor (Red White Matt), which is cultivated from Calico Butterfly Moor (Bluish Base) with its scale in red and white. Compared with Hard-Scale goldfish, the texture of Soft-Scale goldfish feels like the iced ross quartz of Shoushan (a famous mountain producing ornamental stones). The color of red here is called "Fei" (scarlet red color in Chinese), while the white is called "Di" meaning ground. The one in the picture shows a perfect matching of red and white, very dynamic and vivid.

云石龙睛蝶尾

Marble
Butterfly Moor

- **花色类别** ／ Color
  软鳞黑白 ／ Marble

- **品种类别** ／ Species
  文形 ＞ 复合变异类 ＞ 龙睛蝶尾型
  Fantail Goldfish ＞ Compound Variation ＞ Butterfly Moor

- **品种详述** ／ Species Description

　　图例所示为云石龙睛蝶尾，是软鳞黑白色的龙睛蝶尾。云石花色和水墨花色都是由软鳞五花选育而出，是近几年来新问世的软鳞花色品种，也颇受市场好评。图例所示龙睛蝶尾体形丰满，尾鳍优美，是品质较高的一尾金鱼。

As is showing in the figures, Marble Butterfly Moor is a Soft-Scale black white variety. Marble color and water pattern varieties are all cultivated from Soft-Scale Calico (Bluish Base) variety. It is a new variety and is well received in the market. The one in the picture with a full figure, faery caudal fin, is very outstanding.

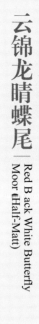

云锦龙睛蝶尾
Red Back White Butterfly Moor (Half-Matt)

花色类别 / Color •

软鳞五花 / Red Black White (Half-Matt)

品种类别 / Species •

文形 > 复合变异类 > 龙睛蝶尾型
Fantail Goldfish > Compound Variation > Butterfly Moor

品种详述 / Species Description •

　　图例所示为云锦龙睛蝶尾，是从荧鳞龙睛蝶尾中选育而出。云锦龙睛蝶尾全身为蓝黑色硬质鳞，边缘为银白色，侧线上部会有如同水印般的淡金黄色花斑，头顶和背脊处有朱砂色红斑。之所以用云锦命名，一是其色如云锦般典雅高贵，二是其鳞片淡金色花斑与银白色鳞片边缘相互辉映，如同云锦中的夹金银线妆花工艺。云锦龙睛蝶尾要求全身为蓝黑色硬鳞，且色彩稳定终生不褪，十分珍稀。

The figures show the Red Black White Butterfly Moor (Half-Matt), it is cultivated from Glimmering Scale Butterfly Moor. With blue black Hard-Scale, white on its edge, and the lateral line is like light golden yellow spot with water ink; on the head and back, there are dark red spots. With the Chinese name "Yunjin", it presents an image of dignity and elegance like the "Yunjin" barcode (Yunjin, special thread for emperor's dragon robe); in addition, the pale gold and silver flake spotted and white scales edge is the same as Yunjin with brocade in gold and silver thread to makeup flower clip. Red Black White Butterfly Moor (Half-Matt) must be blue black and Hard-Scale, with the stable characteristics, it is regarded very rare.

麒麟龙睛蝶尾

Kirin Butterfly Moor

- **花色类别** / Color

  软鳞五花 / Kirin

- **品种类别** / Species

  文形 > 复合变异类 > 龙睛蝶尾型

  Fantail Goldfish > Compound Variation > Butterfly Moor

- **品种详述** / Species Description

  图例所示为麒麟龙睛蝶尾。麒麟色也是近几年来颇为流行的软鳞花色品种，由软鳞五花色选育而出。其全身墨蓝色，鳞片基部色较深，边缘色浅，全身有黑点碎花。全身墨蓝色的称之为墨麒麟，有红色斑块的称之为花麒麟。图例所示为花麒麟，身形娇俏，尾形优美，背鳍处红斑若移至头顶则更佳。

  The figure shows a Kirin Butterfly Moor. It belongs to Soft-Scale category cultivated from Soft-Scale Calico (Bluish Base) Goldfish, and is well received in the market. It is black blue all over the body, dark on the fin root, light on the fin edge with black spots scattered. The ones with black green are called Black Kirin Butterfly Moor; the ones with red spots are called Calico Kirin Butterfly Moor. The one in the picture is Calico Kirin Butterfly Moor with slim body, elegant tail. If the dorsal fin grows on the top of head, it will be the best.

花色类别 / Color •
软鳞五花 / Intense Colors

品种类别 / Species •
文形 > 复合变异类 > 龙睛蝶尾型
Fantail Goldfish > Compound Variation > Butterfly Moor

品种详述 / Species Description •

重彩龙睛蝶尾
Butterfly Moor
(with Intense Colors)

　　图例所示为重彩龙睛蝶尾。重彩龙睛蝶尾也是从荧鳞龙睛蝶尾中选育而出的，其部分鳞片褪白，软鳞化，黑色部分的墨和红色部分的绯所占面积较大，给人较强的视觉冲击力。重彩龙睛蝶尾适合水族箱玩赏，配以适当的灯光，可以将其绚丽的花色淋漓尽致地呈现出来。重彩龙睛蝶尾和其他重彩花色系列的金鱼一样，十分受追捧。

The figures show the Butterfly Moor (with Intense Colors), a breeding selected from Glimmering Scale Butterfly Moor, some of its scales faded to white, and becomes very soft. The pretty black ink and red color with big portion gives a strong visual impact. This variety is better to be appreciated in an aquarium tank, with appropriate lighting, it shows out very dazzling color. Like other intense color varieties, the Butterfly Moor (with Intense Colors) is also very popular.

软鳞五花 / Intense Colors

五花龙睛蝶尾
Calico Butterfly
Moor (Bluish Base)

- **花色类别** / Color
  软鳞五花 / Calico (Bluish Base)

- **品种类别** / Species
  文形 > 复合变异类 > 龙睛蝶尾型
  Fantail Goldfish > Compound Variation > Butterfly Moor

- **品种详述** / Species Description

　　五花龙睛蝶尾是较受欢迎的传统金鱼品种之一，以江苏如皋和苏州出产的翠蓝花色较受市场青睐。近些年来，由于和荧鳞蝶尾过度杂交的缘故，导致其基因受到污染，纯正的五花龙睛蝶尾已经很难寻觅。五花龙睛蝶尾系软鳞花色蝶尾的代表品种，其蓝背、白腹、红顶、黑丝尾配以黑色和橙色碎花最为标准，全身基本无硬质反光鳞片，尾鳍张开如飞舞的蝴蝶，尾鳍先端可及鳃盖甚至眼睛。五花龙睛蝶尾二龄以后饲养应适当减少光照，保持其素蓝色的稳定。

Calico Butterfly Moor (Bluish Base) is one of the most popular traditional goldfish. Rugao and Suzhou are the two places boost for high quality Bright Blue Calico Butterfly Moor (Bluish Base). In recent years, due to too much hybridization with Fluorescent Scale Butterfly Moor, its gene has been contaminated, thus pure Calico Butterfly Moor is rare in the market. Calico Butterfly Moor (Bluish Base) is a classic variety of colorful Soft-Scale butterfly moor, generally with blue back, white belly, red head, black strip-shaped tail decorated with black and orange trims. It has no hard and reflective scales at all. It looks like a butterfly when the caudal fin is unfolded, the top of the caudal fin can reach the gills, even eyes. Less sunshine shall be given to the Calico Butterfly Moor to stable the blue color after two years old.

文形
Fantail Goldfish
∨
复合变异类
Compound Variation
∨
龙睛高头型
Oranda with Dragon Eyes

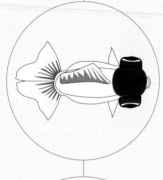

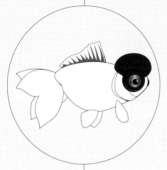

　　龙睛高头又称龙睛帽子，是眼型变异类与头型变异类杂交选育出的一个复合变异类品种，也是传统金鱼的代表品种之一。20世纪80年代，朱砂眼白龙睛高头曾经和众多中国金鱼精品一起，远赴美国纽约展出，引起轰动。大多数金鱼在营养条件充足的情况下，饲养久了，头部都会有增生，即使草金也不例外，因此高头在遗传中也是属于显性遗传。龙睛高头的花色品种也较多，古老的花色品种有红龙睛高头、黑龙睛高头、紫龙睛高头、红白龙睛高头等，近几年又培育出雪青龙睛高头、紫蓝龙睛高头、重彩龙睛高头等等。传统金鱼中属于龙睛高头类的，还有一个最知名的品种，就是龙睛鹤顶红了。龙睛高头多数体质较强，易于饲养，侧视俯视俱佳。龙睛高头中，偶尔还会出现灯泡眼的叠加变异，十分罕见。

Oranda with Dragon Eyes, also called Caped Oranda with Dragon Eyes, is a Compound Variation crossed by Eye Variation and Head Variation. It is one of the traditional varieties. In 1980s, The Cinnabar Oranda with Dragon Eyes together with other goldfish varities, have been brought to America for exhibition, then largely increasing its fame. Most of the goldfish, if with good nutrition will all grow with pompon after a long time, no exception for Common Goldfish. So obviously, pompon is of dominant genes. There are many color species, like the ancient Red Oranda with Dragon Eyes, Black Oranda with Dragon Eyes, Chocolate Oranda with Dragon Eyes, Red White Oranda with Dragon Eyes. Currently, more varieties are developed, such as Lilac Oranda with Dragon Eyes, Chocolate Blue Oranda with Dragon Eyes, and Oranda with Dragon Eyes (with Intense Colors) and so on. Among the traditional Oranda with Dragon Eyes, the most famous breed is Red Caped Oranda with Dragon Eyes. The physique of this breed is very strong so it is very easy to be kept. The best appreciation is to be viewed from the above. Occasionally, the bulb light variety rises, which has become very rare.

花色类别 ／ Color •

红 ／ Red

品种类别 ／ Species •

文形 ＞ 复合变异类 ＞ 龙睛高头型

Fantail Goldfish ＞ Compound Variation ＞ Oranda with Dragon Eyes

品种详述 ／ Species Description •

红龙睛高头 ｜ Red Oranda with Dragon Eyes

　　图例所示为红龙睛高头。其体形雄健，顶茸丰满，双眼凸出且对称，尾鳍宽大而飘逸。其鼻瓣膜有发育但还达不到绒球的标准，如果再不断提纯筛选，应该可以达到龙睛高头球的标准。红龙睛高头属于小众型花色品种，极少有金鱼养殖场饲养，因此较为珍稀。红龙睛高头侧视俯视均可，但偏向于侧视，适合水族箱玩赏。

The figures show the Red Oranda with Dragon Eyes. It is with strong physics, developed head pompon; the eyes are protruding and symmetrical, and the dorsal fin is wide stretching. Nasal valve has developed but not as compact as a pompon, if keep selecting, it could be developed into the same standard as Oranda with Dragon Eyes and Pompons. This variety is kept within a small minority, so it is very precious. It is both ok to appreciate from above or from the side, better from side, and is very appropriate to be kept in the aquarium.

花色类别 / Color •

紫 / Chocolate

品种类别 / Species •

文形 > 复合变异类 > 龙睛高头型
Fantail Goldfish > Compound Variation > Oranda with Dragon Eyes

品种详述 / Species Description •

紫龙睛高头

Chocolate Oranda with Dragon Eyes

　　图例所示为紫龙睛高头，紫龙睛高头是一个老品种，其外部特征和传统的紫帽子几乎完全一样，所不同的是紫帽子平眼，而紫龙睛高头是鼓眼。紫龙睛高头属于小众花色品种，饲养的金鱼养殖场极少，濒临灭绝，是一个需要加以保护的金鱼品种。

The figures show the Chocolate Oranda with Dragon Eyes. It is an old variety, bearing the same appearance as traditional Chocolate Caped Oranda, the only difference is that the first one is with protruding eyes while the later one is with flat eyes. Since this variety is very rare to keep aquarium and almost extinguished, it should be well protected.

雪青龙睛高头

Lilac Oranda with Dragon Eyes

- **花色类别** / Color
  雪青 / Lilac

- **品种类别** / Species
  文形 > 复合变异类 > 龙睛高头型
  Fantail Goldfish > Compound Variation > Oranda with Dragon Eyes

- **品种详述** / Species Description

　　图例所示是雪青龙睛高头。雪青龙睛高头是利用雪青蝶尾和文形金鱼中的虎头型杂交选育出的一个新的花色品种。其顶茸适中，身形饱满，尾鳍宽大飘逸，全身淡雪青色，基本无杂色。雪青龙睛高头适宜在白色陶瓷缸中饲育玩赏，其色苍润古朴，赏之如闻古琴高山流水之音。

The figures show the Lilac Oranda with Drgon Eyes.It is the breeding selected from Lilac Butterfly Moor and Tigerhead in Fantail Goldfish. The pompon is in medium size, entirely lilac with full figure, wide stretching fins. It is appropriate to be kept in the white pottery basin. With the ancient looking of the color, it provides people with anidyllic scenery.

花色类别 / Color •

红白 / Red White

品种类别 / Species •

文形 > 复合变异类 > 龙睛高头型

Fantail Goldfish > Compound Variation > Oranda with Dragon Eyes

品种详述 / Species Description •

龙睛鹤顶红

Red Cap Oranda with Dragon Eyes

　　图例所示为龙睛鹤顶红。龙睛鹤顶红实际就是红白龙睛高头金鱼的一个巧色品种。由于其出现较早，加之深受欢迎，经过金鱼业者不断地提纯筛选，这种花色得以稳定遗传。龙睛鹤顶红其外形和鹤顶红基本相同，唯一区别就是眼睛产生了龙睛型变异。相对于其他龙睛高头，龙睛鹤顶红一直较受欢迎，全国各地金鱼养殖场都有饲养。龙睛鹤顶红中，还有一种顶茸褪色的个体，其顶茸呈淡柠檬黄色，雅号白龙托玉。

As is shown in the figures, Red Cap Oranda with Dragon Eyes is with the delicate color from Red White Oranda with Dragon Eyes. It has been raised for a long time, and well received by people. After continuous selected breeding, this variety is now very stable genetic. The Red Cap Oranda with Dragon Eyes and Red White Oranda with Dragon Eyes is almost same except for the eyes. It has gained a large popularity and been kept throughout the whole country. There is another species, called White Dragon Holding Jade because its head pompon fades to light lemon.

红白龙睛高头

Red White Oranda
with Dragon Eyes

- **花色类别** / Color

  红白 / Red White

- **品种类别** / Species

  文形 > 复合变异类 > 龙睛高头型

  Fantail Goldfish > Compound Variation > Oranda with Dragon Eyes

- **品种详述** / Species Description

  　　图例所示为红白龙睛高头，其外部体态特征与红龙睛高头基本一致，体色为红白花色。红白花色系列的金鱼之所以受到欢迎，就是因为有了变化，而红白花色的变化既不繁杂又很柔和，同时体现出中国传统哲学思想中的阴阳之说。图例1、3所示为红顶红尾的首尾红花色品种，配以朱砂眼和红色的胸鳍，体现出清新典雅之美。

  As is shown in the figures, the Red White Oranda with Dragon Eyes, is the same height as Red Oranda with Dragon Eyes. The reason why it is so popular is that it keeps changing on the color; the color is simple and gentle, reflecting the philosophy of "Yin" and "Yang" (the two opposing principles in nature). The figure 1,3, is a variety red on tail and head, with cinnabar eyes and red chest fin, representing an elegant and pure image.

紫蓝龙睛高头

Chocolate Blue Oranda with Dragon Eyes

- **花色类别** / Color
  紫蓝 / Chocolate Blue

- **品种类别** / Species
  文形 > 复合变异类 > 龙睛高头型
  Fantail Goldfish > Compound Variation > Oranda with Dragon Eyes

- **品种详述** / Species Description

　　图例所示为紫蓝龙睛高头，和雪青龙睛高头一样，也是利用雪青蝶尾和文形金鱼中的虎头型杂交选育出的一个新的花色品种。纯色无紫斑的为雪青，带紫斑的即是紫蓝花色。紫蓝花色和雪青花色是同胞兄弟，相伴相生。紫蓝花色的紫斑如果分布巧色，就会产生十二紫的花色，十分罕有。

The figures show the Chocolate Blue Oranda with Dragon Eyes. As same height as Lilac Oranda with Dragon Eyes, it is a new breed crossed by Lilac Butterfly Moor and Tigerhead in Fantail Goldfish. Pure without chocolate spot is lilac, with chocolate spot is chocolate blue. Chocolate blue and Lilac are twin brothers. Once the chocolate blue shows ingenious color on chocolate spot, the twelve chocolate will appear, that is very rare.

文形
Fantail Goldfish

ⓥ

复合变异类
Compound Variation

ⓥ

龙睛球型
Telescope with Pompons

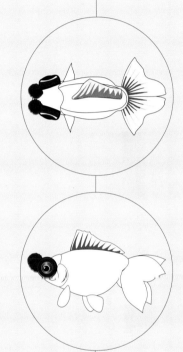

　　龙睛球也是出现较早的一个复合变异类金鱼品种，是龙睛与文球杂交选育而成，目前可以看到的记述，其出现在1935年左右。龙睛球球体紧实，一般双球较常见，四球较罕见。其游动时，绒球不停颤动，极富动感。龙睛球有时会伴有顶茸增生，尤其二三龄以后较为明显。龙睛球花色也较丰富，目前可以看到的有红龙睛球、黑龙睛球、蓝龙睛球、紫龙睛球、雪青龙睛球、红白龙睛球、黑白龙睛球、紫蓝龙睛球、三色龙睛球、五花龙睛球、樱花龙睛球和麒麟龙睛球等等。龙睛球饲养时，在一龄时可以较高密度饲养，控制其生长速度，俗称"压鱼"，经过这个过程的龙睛球尾大球紧、身材匀称。龙睛球性格活泼，较易与人亲近。其体质强健，摄食能力强，易于饲养，且俯视和侧视都有较好的观赏效果。

Telescope with Pompons shows up very early as the compound variation variety, hybridized by the Dragon Eye and Fantail Goldfish. It is recorded to appear in 1935. The head pompon of this variety is very compact and tight, normally with two pompons. It's very rare to see four head pompons. When swimming, the pompon vibrates. The Pompon can be proliferated after two or three year old. The colors at present are Red Telescope with Pompons, Black Telescope with Pompons, Blue Telescope with Pompons, Chocolate Telescope with Pompons, Lilac Telescope with Pompons, Red White Telescope with Pompons, Black White Telescope with Pompons, Chocolate Blue Chocolate Telescope with Pompons, Red White Black Telescope with Pompons, Calico Telescope with Pompons (Bluish Base), Sakura Telescope with Pompons (Red White Matt) and Kirin Telescope with Pompons, etc. This category on the first year, can be intense bred to control the growth speed. This is so called the pressure of goldfish. Telescope with Pompons after this process normally grows with large compact head pompon and is moderate proportioned. This kind of variety is very active, easy to get close to people. With strong physics and good digestion, it is very easy to be raised. It is best to be appreciated by viewing from above or from the side.

红龙睛球

Red Telescope with Pompons

- **花色类别** / Color

  红 / Red

- **品种类别** / Species

  文形 > 复合变异类 > 龙睛球型

  Fantail Goldfish > Compound Variation > Telescope with Pompons

- **品种详述** / Species Description

  图例所示为红龙睛球，大约出现在1935年，是龙睛球系列中最早问世的花色品种。图例所示红龙睛球球体紧实，双眼对称，身形匀称，头身比例适度，尾鳍舒展飘逸。为了保持双球的紧实对称，可以对球体进行适当修剪。红龙睛球饲养时操作要小心，一旦球体受损，需要较长时间才能恢复。

  The figures show a Red Telescope with Pompons, which appeared in 1935, is earliest variety of Telescope with Pompons. The one here has a tight compact pompon, symmetrical eyes and good figure size. The head and the body is well proportioned, and the caudal fin is wide and stretching. In order to retain the pompon in good condition, appropriate pruning might be needed. Just be careful, once the pompon gets hurt, it needs long time to recover.

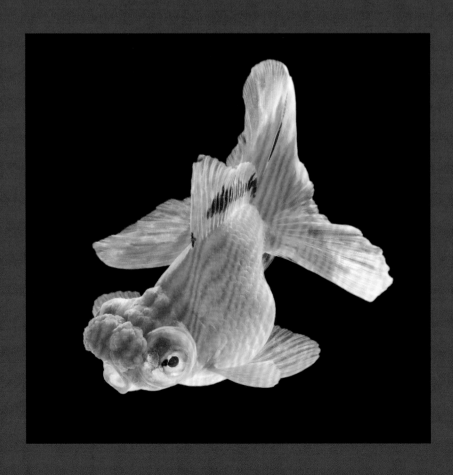

白龙晴球 — White Telescope with Pompons

花色类别 / Color •

白 / White

品种类别 / Species •

文形 ＞ 复合变异类 ＞ 龙晴球型

Fantail Goldfish ＞ Compound Variation ＞ Telescope with Pompons

品种详述 / Species Description •

　　图例所示为白龙晴球，由红白龙晴球选育而出，一般红白花色金鱼中出现的纯白色个体称为"白皮"或"白条"，很少予以保留。图例所示白龙晴球眼球内有虹彩，和白色身体以及淡柠檬黄的绒球搭配，显得特别素净高洁，也十分漂亮。若是类似龙晴蝶尾中玛瑙眼的个体，这样的花色则更吸引眼球。

As is shown in the figures, White Telescope with Pompons is developed from Red White Telescope with Pompons. Generally, the pure white variety is called White Skin or White Stick, which is very rare to see. The one in the picture, with the rainbow color in the eyes, white figure and light lemon pompon, is so elegant and graceful. It will be perfect if the eye balls are in agate color.

黑龙睛球
—Black Telescope with Pompons

- **花色类别** / Color

  黑 / Black

- **品种类别** / Species

  文形 > 复合变异类 > 龙睛球型

  Fantail Goldfish > Compound Variation > Telescope with Pompons

- **品种详述** / Species Description

  图例所示为黑龙睛球。黑龙睛球外部形态与红龙睛球完全一样，其体色为黑色，二三龄后会向红色转变成为红龙睛球。黑龙睛球中会出现全身墨黑而唯独两个绒球呈现红色的名贵花色"朱球墨龙睛"，极其美丽。但这类巧色很难保持。目前可以通过紫龙睛红球和蓝龙睛球杂交，其子代中会出现类似朱球墨龙睛的个体，相对于黑龙睛球中出的要稳定许多。

As is shown in figures, it is a Black Telescope with Pompons, whose eyes look exactly the same as Red Telescope with Pompons. The body color is black, normally turns red after two to three years later, and becomes the Red Telescope with Pompons. The Black Telescope can be varied as jet black in entire body but only red on the two Pompons, this is called Black Telescope with Red Pompons. However, breeding like this is very hard to retain. At present, it can be developed by crossing the Chocolate Telescope with Pompons and Blue Telescope with Pompons, more stable than the ones developed from Black Telescope with Pompons.

蓝龙晴球 —— Blue Telescope with Pompons

- **花色类别** / Color

  蓝 / Blue

- **品种类别** / Species

  文形 > 复合变异类 > 龙晴球型

  Fantail Goldfish > Compound Variation > Telescope with Pompons

- **品种详述** / Species Description

  　　图例所示为蓝龙晴球。蓝龙晴球也是传统品种，大约在1953年被培育出来。蓝龙晴球外部形态和红龙晴球基本相同，尾鳍宽大，球体紧实。蓝龙晴球会有部分褪色为黑白色或三色，其中身体褪为黑白色或喜鹊花色，而两个球呈现红色或粉红色的个体，雅称为"喜鹊登梅"，暗合喜上眉梢的含义。蓝龙晴球与紫龙晴红球杂交，可以培育出朱球墨龙晴。

  As is shown in the figures, the Blue Telescope with Pompons also belongs to a traditional variety, appeared in about 1953. Like the Red Telescope with Pompons, it has a wide caudal fin and compact pompon. Sometimes, the pompon will fade to black white or tri-color (blue, black and white). For those with its body in black white or Magpie color, the pompon might be changed to red or pink, so is called as Magpie on the plum blossom, meaning happy and lucky. Crossed with Chocolate Telescope with Pompons, the Red Black Telescope with Pompons can be developed.

花色类别 / Color •

紫 / Chocolate

品种类别 / Species •

文形 > 复合变异类 > 龙晴球型

Fantail Goldfish > Compound Variation > Telescope with Pompons

品种详述 / Species Description •

紫龙晴球

Chocolate Telescope with Pompons

　　图例所示为紫龙晴球。紫龙晴球的出现，略早于蓝龙晴球3年，是同时期培育出的花色品种。紫龙晴球外部形态特征同其他龙晴球基本类似，但体形会略小些。紫龙晴球除了纯紫的花色外，还会变化出紫身红球、紫身白球、紫身鸳鸯球等花色。紫龙晴球褪色比例不算太高，饲养较容易。

As is shown in the figures, the Chocolate Telescope with Pompons showed up 3 years earlier than Blue Telescope with Pompons, they are the contemporary varieties of the same category. Its pompon looks similar to other Telescope with Pompons varieties, but a little smaller. Except Chocolate, there are also Chocolate Telescope with Red Pompons, or with White Pompons, or with two different colors on each side. Chocolate Telescope with Pompons is not likely to get fading, so it is comparatively easy to be kept.

花色类别／Color •

红白／Red White

品种类别／Species •

文形 ＞ 复合变异类 ＞ 龙睛球型
Fantail Goldfish ＞ Compound Variation ＞ Telescope with Pompons

品种详述／Species Description •

红白龙睛球
Red White Telescope with Pompons

　　图例所示为红白龙睛球。红白龙睛球可以有许多巧色花色，如全身洁白，唯独两个红色绒球的朱砂龙睛球；或是全身洁白，两个眼球全红的玛瑙眼龙睛球；还有二者花色相结合的玛瑙眼朱砂龙睛球等等。饲养红白龙睛球时最好在一龄鱼时适当"压鱼"，经过这个过程的龙睛球身材匀称、球紧尾大、颜色浓郁，较符合优秀龙睛球的标准。

As is shown the figures, the Red White Telescope with Pompons gets many color categories, such as white all over the body with the two red pompons, or with the agate eyeballs, or the compound Agate Red Telescope with Pompons. Red White Telescope with Pompons should be pressed during the first year, ensuring the fine figure and intense and compact pompon. It is aligned with the appreciation standard.

红黑龙睛球
— Red Black Telescope with Pompons

花色类别 / Color •

红黑 / Red Black

品种类别 / Species •

文形 > 复合变异类 > 龙睛球型
Fantail Goldfish > Compound Variation > Telescope with Pompons

品种详述 / Species Description •

　　红黑龙睛球是黑龙睛球向红龙睛球转色的过渡阶段品种，其最终会转色成红龙睛球。图例所示龙睛球双红球紧实，体态优美，背鳍高耸挺立，尾鳍宽大飘逸。饲养时，可以采用降低水温等方式，延缓其褪色过程，达到最佳欣赏效果。

The Red Black Telescope with Pompons is the transit variety from Black Telescope with Pompons to Red Telescope with Pompons, finally it will change to Red Telescope with Pompons. As is shown here, this telescope is with two compact red pompons, elegant figure, straight dorsal fin, and wide caudal fin. The fading of its color can be controlled by lowering the water temperature so as to reach the best appreciation effect.

文形
Fantail Goldfish

∨

多重复合变异类
Compound Variation

∨

龙睛皇冠珍珠型
Crown Pearlscale with Dragon Eyes

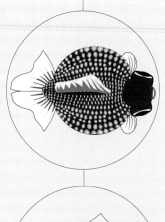

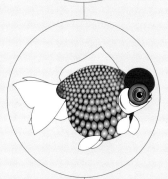

　　龙睛皇冠珍珠是皇冠珍珠中产生龙睛型变异的突变个体，金鱼业者对其提纯筛选加以固化而培育出的一个品种类型。龙睛皇冠珍珠顶茸发达的不多，多呈双瓣鸡心形。龙睛皇冠珍珠市场接受度不高，所以一直是个小众品种，饲育的金鱼场也不多，多由皇冠珍珠中变异个体获得，单独育种较少。在皮球珍珠中，也会产生龙睛型变异的突变个体，称为鼓眼珍珠，以前国内较常见，而现在几乎已经绝迹。鼓眼皮球珍珠在东南亚的金鱼养殖场还有饲养，但风格和国内原来的不太一样。龙睛皇冠珍珠的花色不多，有黑色、红白色、五花色，其体质较强，较易饲养。

Crown Pearlscale with Dragon Eyes is selected and purified by goldfish raiser from Dragon Eyes variation from Crown Pearlscale. The head pompon is twin hearted, but not so developed. It is not well received in the market, so is only kept in a small group and either so common in goldfish aquarium. Mostly, it is bred from Crown Pearlscale verity, not from separate cultivation. Among the Golfball Pearlscale, a variation called Protruding Eye Pearlscale can also be produced, seen frequently in China in the past, now almost extinct. But they are still kept in South Asia, with a different style from China. The color category is not so rich, consisted of Black,Red white, calico. It has strong physics so it is easy to be raised.

花色类别 / Color •

黑 / Black

品种类别 / Species •

文形 > 多重复合变异类 > 龙睛皇冠珍珠型

Fantail Goldfish > Compound Variation > Crown Pearlscale with Dragon Eyes

品种详述 / Species Description •

黑龙睛皇冠珍珠 — Black Crown Pearlscale with Dragon Eyes

图例所示为黑龙睛皇冠珍珠。其全身漆黑如墨，珍珠鳞饱满，排列规整。顶茸双瓣，发育良好。背鳍和腹鳍有残损，在操作时要当心，鱼鳍的自愈性虽然很强，但还是要做到谨慎操作。黑龙睛皇冠珍珠需要有充足的光照，降低水温及充足的日照可以促使它愈发浓黑。黑龙睛皇冠珍珠适应性强，体质健壮，较易饲养。

As is shown in the figures, it is jet black all over the whole body. The scales are full and organized; the two head pompons are developed. The dorsal fin and pelvie fin is damaged; although it is easy to get recovered, still needs to be careful for operation. Black Crown Pearlscale with Dragon Eyes needs to have plenty sunshine. The cool water and enough sunshine can make it much blacker. It has a strong adaptability, so it is easy to be kept.

花龙睛皇冠珍珠 | Calico Crown Pearlscale with Dragon Eyes

- **花色类别** / Color
  珍珠鳞五花 / Calico

- **品种类别** / Species
  文形 > 多重复合变异类 > 龙睛皇冠珍珠型
  Fantail Goldfish > Compound Variation > Crown Pearlscale with Dragon Eyes

- **品种详述** / Species Description

　　图例所示为花龙睛皇冠珍珠亚成体。花龙睛皇冠珍珠是从五花皇冠珍珠培育出的花色品种，全身素洁，点缀有黑色斑点。其顶茸发育适中，呈淡柠檬黄色，身形娇小，腹部珍珠鳞有残缺，是操作不慎造成的。珍珠鳞可以修复，但比其他普通鳞片需要的时间长许多。因此对此类金鱼的操作一定要谨慎。

As is shown in the figures,Calico Crown Pearlscale with Dragon Eyes is a sub-adult. It is developed from Calico Crown Pearlscale with Dragon Eyes (Bluish Base), scattered with black spots on the body. The head pompon is well developed in lemon; the figure is quite small with incomplete scales on the belly which might be caused by inappropriate operation. Recovering takes a long time so it must be very careful to operate.

五花龙睛皇冠珍珠

Calico Crown Pearlscale with Dragon Eyes (Bluish Base)

- **花色类别** / Color
  珍珠鳞五花 / Calico (Bluish Base)

- **品种类别** / Species
  文形 > 多重复合变异类 > 龙睛皇冠珍珠型
  Fantail Goldfish > Compound Variation > Crown Pearlscale with Dragon Eyes

- **品种详述** / Species Description

　　图例所示为五花龙睛皇冠珍珠亚成体，是五花皇冠珍珠中产生龙睛型变异的突变个体。五花龙睛皇冠珍珠以蓝色基底为贵，而图例所示金鱼其红色斑块太多太大。其身形灵巧，尾鳍轻盈，游动稳健。五花龙睛皇冠在龙睛皇冠珍珠中占据的比例最高，可见大家对这种花色的喜爱。该类型金鱼生长过程中要消耗大量的钙磷等无机盐，如果不及时补充，则会引起缺钙，其典型症状为鳃盖翻卷，珍珠鳞不饱满甚至变为普通的平鳞，影响观赏。

The figures show the sub-adult of Calico Crown Pearlscale with Dragon Eyes (Bluish Base), is variation of Calico Crown Pearl with Dragon Eyes(Bluish Base). Calico Crown Pearlscale with Dragon Eyes (Bluish Base) is much valuable, however the one shown is with too much red patches. The upper body is dexterity, tail is light, and swims sound. There is a large portion for Calico Crown Pearlscale with Dragon Eyes (Bluish Base) taking up in the Crown Pearscale with Dragon Eyes variety, this shows how much people like it. During the growth, a large amount of calcium and phosphorus will be consumed, if not supplemented on time, can lead to calcium deficiency. The typical symptoms are gill cover rolling, cone scales not full even become regular flat scales, which largely affects the appreciation value.

蛋形
Egg-Fish

中国金鱼从外部形态上可以区分为两大类别，即文形和蛋形，这是中国金鱼分类树的两大基本分支。如同中国太极图中阴阳两尾鱼生化为万物一样，蛋形金鱼和文形金鱼共同演化出多姿多彩的中国金鱼世界。

和文形金鱼一样，蛋形金鱼的名称也是由其原型类品种蛋鱼而来。蛋鱼的特征有：头小而尖，背鳍宗全退化消失，其他鱼鳍也比文形金鱼更为短圆，身体浑圆似蛋，故而得名。所有背鳍退化消失并具有稳定遗传的金鱼，全部归为蛋形金鱼。

从现有的资料考证，蛋形金鱼的出现较文形金鱼要晚许多。明代的图刻本中有记载一种无背鳍的草金，名曰金兰子。而蛋鱼究竟是从金兰子演化过来的，还是由文鱼中背鳍退化的个体，经过漫长的人工选育而逐步演化而来的，目前已无法考证。而金兰子这个品种也早已湮灭在历史的长河之中了。

蛋形金鱼覆盖的品种数略少于文形金鱼。以蛋形金鱼为纲，其下可划分出原型类、头型变异类、眼型变异类、鼻膜变异类和复合变异类。蛋形金鱼对背部要求严格，背鳍消失的部分必须平滑光顺，无鼓钉无残鳍，因此其淘汰率远高于文形金鱼。从金鱼的演化角度而言，蛋形金鱼由于背鳍消失，相对其始祖金鲫鱼而言，在形态上差异更大，因而，在同一变异类型的基础上，蛋形金鱼的演化要高于文形金鱼，也更受金鱼饲养者的喜爱。

Chinese goldfish from the external morphology can be divided into two categories, namely, Fantail Goldfish and Egg-Fish, which are known as two basic branches of Chinese Goldfish classification. Just like Chinese Taiji Figure of yin and Yang, two fish, make all lives on earth, Egg-Fish and Fantail Goldfish together have developed a colorful Chinese goldfish world.

Like Fantail Goldfish, the name of the egg-shaped goldfish also developed from original variety Egg-Fish. Egg-Fish, from the external morphology, small pointed head, dorsal fin degenerated and completely disappeared. Compared with other fins of Fantail Goldfish, the fins of Egg-Fish are shorter and rounder. Egg-Fish is named after its Egg-Fish body which is just like an egg. All the goldfish types with degreased disappeared dorsal fin and stable inheritance are classified as Egg-Fish.

According to the existing information research, Egg-Fish appears much later than Fantail Goldfish. From Ming Dynasty edition there are records of a non-dorsal fin grass golden fish, called Jin Lanzi. While, what is the Egg-Fish from, Jin Lanzi, or by gradually evolving from Fantail Goldfish degradation dorsal fin individuals after a long artificial breeding, has been unable to research. And the varieties of Jin Lanzi has also annihilation in the long river of history.

The varieties of Egg-Fish is slightly less than Fantail Goldfish. Egg-Fish is taken as the ancestral link of today' Original Type, Head Variation, Eye Variation, Nasal Variation and Compound Variation. High quality Egg-Fish will display a smooth arch-shaped back, no drum nail and residual fins, therefore its elimination rate is far higher than Fantail Goldfish. From the goldfish evolution point of view, compared with the ancestor of golden carp, the dorsal fin disappearance of Egg-Fish make much more morphological differences. Therefore, on the basis of the same variation type, the evolution of Egg-Fish is higher than that of the Fantail Goldfish. Hence Egg-Fish is more popular by goldfish breeders than Fantail Goldfish.

蛋形
Egg-Fish
∨
原型类
Original Type
∨
蛋鱼型
Egg-Fish

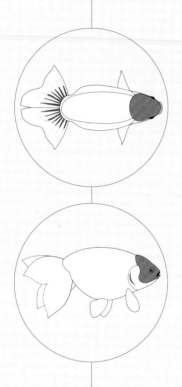

　　蛋形金鱼的原型类就是蛋鱼，也称蛋金。蛋形金鱼的名称也是由此而来。由于金鱼进化的脚步非人力可以阻止，而且相较其他金鱼品种而言，蛋鱼观赏点不多，颜值较低，因此，和文鱼一样，目前我们所看到的蛋鱼，已非原来的纯正意义上的蛋鱼，多为通过杂交获得的一个品种类型。但是，蛋鱼的出现，在整个金鱼演化史中具有里程碑式的重大意义，它的出现开创了一个新的金鱼品系大类。目前中国金鱼里，蛋鱼主要为各色短尾蛋金和长尾蛋金两大类。日本的金鱼出云南京也可划入这个类别当中。蛋鱼在很早前就经福建、广东一带传入日本，由于当时以南方的口音称之为"卵虫"，所以在日本一直沿用这个名称并逐步将它培育成日本的金鱼之王——日寿。

The Original Type of the Egg shaped gold fish is Egg-Fish, is as Egg-Goldfish, from which the name Egg-Fish is derived. The evolution pace of goldfish is beyond people's control. Besides, compared to other varieties of goldfish, the appreciate value of Egg-Fish is lower and it is not so good look as other varieties. Therefore, just like Fantail Goldfish, the Egg-Fish has been non-original genuine. Most of them are varieties by cross hybridization. However, The occurrence of Egg-Fish has a landmark significance in whole evolution history of goldfish and has created a new line of goldfish categories. Currently, in China, there are two classifications the Egg-Fish fall under due to the various lengths of their tails; the short-tail ('Egg-Fish') and long-tail ("Egg-Fish") . Japanese Izumo Nankin can be also included into Egg-Fish category. Egg-Fish domesticated in Japan in earlier times from Guangdong and Fujian. Because at that time Chinese southern accent called "Luanchong", the name "Luanchong" has been kept from that time and be gradually developed into the king of Japanese goldfish named.

花色类别 ／ Color •

软鳞五花 ／ Calico (Bluish Base)

品种类别 ／ Species •

蛋形 ＞ 原型类 ＞ 蛋鱼型
Egg-Fish ＞ Original Type ＞ Egg-Fish

品种详述 ／ Species Description •

五花蛋鱼——
Calico Egg-Fish
(Bluish Base)

　　图例所示为五花蛋鱼，又称花蛋。其花色是较标准的五花花色，红顶、白腹、素蓝花芝麻点。其尾型属于中短尾，若为短尾则更佳。五花蛋鱼体质较强健，活跃度高，较易饲养，可以与大多数其他种类金鱼混养，但不宜和珍珠、水泡等混养。五花蛋鱼不仅欣赏价值较高，在金鱼杂交育种方面，也是非常具有价值的亲本素材。

As shown in the illustration is Calico Egg-Fish (Bluish Base) also known as Calico Egg-Fish. Its color is more standard Calico color: red roof, white-bellied, plain blue flower with sesame point. It has middle-short tail. If shorter, it is better. Calico Egg-Fish with strong physique, high activity is easy to rear. It could be mixed breed with most other types of goldfish with exception for pearls and Bubble-Eye. Calico Egg-Fish (Bluish Base) is not only in high appreciate value, but also a very valuable material in terms of cross-breeding.

蛋形
Egg-Fish

∨

原型类
Original Type

∨

蛋凤型
Egg-Fish Phoenix Tail

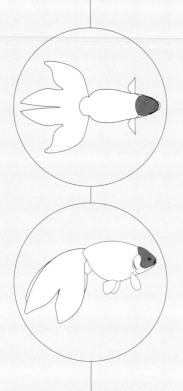

　　蛋鱼根据尾型的长短可以划分为两类，短尾的称之为蛋鱼或蛋金，长尾的称之为蛋凤。标准的蛋凤头部较尖，身体狭长，鱼鳍较一般蛋鱼要长，尤其是尾鳍，长如飘带状，尾叶也较窄，二龄以后尾叶自然下垂，很像是传说中凤凰的尾羽一般，故名之凤尾蛋鱼，简称蛋凤。蛋凤中，有一种全身雪白，唯独头顶有一鲜红色色块的花色品种，其花色如同初升的红日，雅称丹凤朝阳，又称丹凤。蛋凤与丹凤这两个名称因读音很接近，有时容易混淆，应注意区分。与蛋鱼一样，目前可见的蛋凤多为现代杂交选育而出，而非原来纯正的蛋凤。比起古老蛋凤，现代蛋凤大多头部略宽，背部较平顺，尾鳍也更宽大，体形较雄健。饲养蛋凤可以适当增加密度，体形贵瘦不贵腴，太丰满则会失去蛋凤的韵味。

There are two classifications of Egg-Fish by the length of the tail. The short-tailed called Egg-Fish and the long-tailed called Egg-Fish Phoenix Tail. Standard Egg-Fish Phoenix Tail has pointed head, narrow body, long fins which are normally longer than that of Egg-Fish. Especially the caudal fin is as long as a ribbon. Its tail leaf is relatively narrow which will droop naturally after 2 years later, just like the tail of legendary phoenix. So it is named Egg-Fish Phoenix Tail, and people call it phoenix for short. There is one type of goldfish in Egg-Fish Phoenix Tail family which is totally iridescent white except for a bright stain on its head, just like the rising sun, officially named Red phoenix rising sun, also named Red phoenix. In Chinese, the pronunciations of Egg-Fish Phoenix Tail (Dan Feng in Chinese) and Red phoenix, (Dan Feng in Chinese) are very close to and it is very easy to be confused. So it is necessary to distinguish them. Just as Egg-Fish, most of the existing Egg-Fish Phoenix Tail are not the original pure variety but by modern selected rearing. Compared to the previous ancient Egg-Fish Phoenix Tail, modern Egg-Fish Phoenix Tail has slightly wider head, relatively smoother back, a little bigger caudal fin, and a more vigorous body. Appropriately increase the food when feeding it. Slimmer is better than fatter, because it will lose its flavor when it is too fat.

花色类别 / Color •

红 / Red

品种类别 / Species •

蛋形 > 原型类 > 蛋凤型

Egg-Fish > Original Type > Egg-Fish Phoenix Tail

品种详述 / Species Description •

红蛋凤
——
Red Egg-Fish
Phoenix Tail

图例所示为红蛋凤。红蛋凤以前较为常见，现在已经很少有人饲养了。传统的红蛋凤一龄时多为青黑色，转色较晚，转色期多在二龄以后才开始，首先腹部及面部逐渐转为金黄色，而后整个身体再至所有鱼鳍慢慢转为金黄色，最后再由金黄色转为红色。由于转色期晚，鳞片中色素沉积比较厚实，因此颜色显得浑厚并具光泽，红得透彻。优质的红蛋凤从头至尾梢都没有杂色，如同夏天傍晚的火烧云一般。

As shown in the illustration is Red Egg-Fish Phoenix Tail. Red Egg-Fish Phoenix Tail is more common in the past, but now it is rarely reared. Most traditional Red Egg-Fish Phoenix Tail is lividity when it is one year old and normally begins to change its color after it is two. Its belly and face gradually turn into golden first and then the whole body and all fins slowly go into golden, and finally golden to red. Because of veraising late, the flake pigment is deposited relatively thick, the color looks full and shiny, thoroughly red. High quality Red Egg-Fish Phoenix Tail is non-variegated as evening glow in summer.

蓝蛋凤

Blue Egg-Fish
Phoenix Tail

- **花色类别** / Color
  蓝 / Blue

- **品种类别** / Species
  蛋形 > 原型类 > 蛋凤型
  Egg-Fish > Original Type > Egg-Fish Phoenix Tail

- **品种详述** / Species Description

　　蓝蛋凤一直是名贵品种，也是北京金鱼的一个代表品种，曾经作为中日友好交流的珍贵礼物流传到日本，目前日本还有在小范围内饲育蓝蛋凤的原始品种。传统的蓝蛋凤尖头、平背，身体和鱼鳍狭长，全身银灰色鳞片并带有青蓝色的金属光泽。图例所示的蓝蛋凤和传统的品种略有差异，是根据现代金鱼的审美视角加以培育的。其背部加大了曲线的弧度，尾鳍也变得宽大。二者间的审美意趣各有千秋。

Blue Egg-Fish Phoenix Tail is always a rare variety which is also a representative variety of Beijing goldfish. It was once sent to Japan as precious gift for friendly exchanges between China and Japan. Currently, in Japan there are still original Blue Egg-Fish Phoenix Tail reared in a small scale. Traditional Blue Egg-Fish Phoenix Tail's body with a pointed head, flat back, the body and fins are long and narrow. It has silver-colored scales with a green and blue tint. Blue Egg-Fish Phoenix Tail shown in the illustration is a little bit different with traditional varieties which has been artificially reared as the aesthetic perspective of modern goldfish. Curvature has been increased on the back and caudal fin has become bigger. Both of them have their own aesthetic charm.

紫蛋凤 ︱ Chocolate Egg-Fish Phoenix Tail

花色类别 / Color •

紫 / Chocolate

品种类别 / Species •

蛋形 > 原型类 > 蛋凤型

Egg-Fish > Original Type > Egg-Fish Phoenix Tail

品种详述 / Species Description •

　　紫蛋凤和蓝蛋凤一样，在蛋凤金鱼中属于较名贵的花色品种。图例所示紫蛋凤应为二龄左右的成体，各方面的发育度都较好，鱼鳍有点翻折的现象。紫蛋凤的体色根据水质状况和金鱼个体差异，呈古铜色或红铜色，有部分紫蛋凤会转色成红蛋凤。饲养紫蛋凤时要注意不要有尖锐物体刺划到鳞片，紫色鳞片被硬物刺划会留下痕迹，修复的时间比较漫长。饲料部分适当增加矿物质和维生素会使紫蛋凤的鳞片更加有光泽。

Both Chocolate Egg-Fish Phoenix Tail and Blue Egg-Fish Phoenix Tail belong to the rare type of multicolored goldfish. The picture shows a Chocolate Egg-Fish Phoenix Tail roughly two years of age with good growth proportions and its fins have a slight droopy appearance. The color of Chocolate Egg-Fish Phoenix Tail's body can vary between bronze and copper depending on water quality and natural differences. Part of Chocolate Egg-Fish Phoenix Tail will develop to Red Egg-Fish Phoenix Tail. It is important to keep them safe from sharp or pointy objects that might damage their scales and cause scars which will take a long time to heal. Proper adding minerals and vitamins in feeding stuff helps the Chocolate Egg-Fish Phoenix Tail develop lustrous colors.

红白蛋凤

Red White Egg-Fish
Phoenix Tail

- **花色类别** / Color
  红白 / Red White

- **品种类别** / Species
  蛋形 > 原型类 > 蛋凤型
  Egg-Fish > Original Type > Egg-Fish Phoenix Tail

- **品种详述** / Species Description

  红白蛋凤也是蛋凤中较常见的一个花色品种，其中最为名贵的就是全身洁白无杂色，唯独头顶正中有一块红色色斑的巧色金鱼，雅称丹凤朝阳，又称丹凤，民间也称之为戳子红。红白蛋凤中，还有整个头部都是红色的称齐鳃红或元宝红，都属于比较名贵的花色品种。图例所示为普通花色的红白蛋凤。当岁的红白蛋凤尾鳍不是很长，而二龄或三龄以后，尾鳍逐步变长进入最佳观赏期。

  Red White Egg-Fish Phoenix Tail is also a common variety of Egg-Fish Phoenix Tail family. The most rare smart color goldfish has totally white body non-variegated except for a red stain on the center top of the head. it officially named Red phoenix rising sun, also named Red phoenix, and is also called red stamp among people. There is a type of Red White Egg-Fish Phoenix Tail which has a totally red head and it is a relatively rare variety. As shown in the illustration is common colored Red White Egg-Fish Phoenix Tail. One-year-old Red White Egg-Fish Phoenix Tail's caudal fin is not very long while, after it is 2 or 3, caudal fin will grows longer and longer. At that time it is the best period to view.

花色类别 / Color •

红黑 / Red Black

品种类别 / Species •

蛋形 > 原型类 > 蛋凤型
Egg-Fish > Original Type > Egg-Fish Phoenix Tail

品种详述 / Species Description •

红黑蛋凤
—— Red Black Egg-Fish
Phoenix Tail

　　红黑蛋凤，又称包金色蛋凤，大多是红蛋凤转色过程中的一个过渡期品种。由于红蛋凤转色较晚，转色过程有时较长，因此出现红黑蛋凤，适当控制饲料成分配比以及水质和温度，可以延长红黑蛋凤的观赏期。从理论上说，红黑蛋凤最终还是会转变成红蛋凤，这也是金鱼色彩变化的一个魅力所在。红黑蛋凤和红蛋凤一样，较易饲养，适当增加饲养密度，控制好体形不要过于丰腴是养好蛋凤金鱼的关键。

Red black Egg-Fish phoenix, also known as gilded Egg-Fish phoenix Tail. Most of them are the a transition period variety during Red Egg-Fish phoenix Tail going through a color change. The red Egg-Fish phoenix changes color relatively late and the process of changing color takes longer than usual. So it is what brings out the Red Black Egg-Fish Phoenix Tail. The view period can be prolonged by proper feeding as well as the correct water quality and water temperature. In theory, Red Black Egg-Fish Phoenix Tail will eventually change back to Red Egg-Fish Phoenix Tail. Both the Red Egg-Fish phoenix Tail and the Red black Egg-Fish phoenix are fairly easy to keep and rear, the key is to not let them grow too big and fat.

樱花蛋凤
— Sakura Egg-Fish Phoenix Tail (Red White Matt)

- **花色类别** / Color

  软鳞红白 / Red White Matt

- **品种类别** / Species

  蛋形 > 原型类 > 蛋凤型

  Egg-Fish > Original Type > Egg-Fish Phoenix Tail

- **品种详述** / Species Description

  樱花蛋凤是蛋凤中软鳞红白里的一个花色品种。图例所示为樱花蛋凤的亚成体，整个身形包括背部平顺度都比较好，和其他蛋凤金鱼一样，其尾鳍发育要到二龄甚至三龄以后。樱花蛋凤体质较好，较易饲养，其饲养关键同样也是需要控制好体形发育。

  Sakura Egg-Fish Phoenix Tail (Red White Matt) is a soft scale red and white variety in Egg-Fish Phoenix Tail family. The picture shows a juvenile cherry Egg-Fish Phoenix Tail. The whole body including the back are flat and smooth. Its caudal fin will continue to grow until age two, sometimes even past age three. The cherry Egg-Fish Phoenix Tail has a strong physique and is easy to rear as long as it doesn't grow too big.

花色类别 / Color •

软鳞红黑 / Tiger Banded

品种类别 / Species •

蛋形 ＞ 原型类 ＞ 蛋凤型

Egg-Fish ＞ Original Type ＞ Egg-Fish Phoenix Tail

品种详述 / Species Description •

　　虎纹蛋凤也是蛋凤中软鳞五花里的一个花色品种。以前虎纹花色是不被业界认同的，大多作为次品淘汰，但现在已经得到部分认同。图例所示为虎纹蛋凤的亚成体，但其尾鳍发育已经比较成熟，初具成体后的感觉，背部的平顺度尚可接受。花色上，其身上的黑色条纹略显偏少，如果鱼鳍上再点缀些黑色条纹则会更加完美。

Tiger Banded Egg-Fish Phoenix Tail is a Calico (Bluish Base) type in Egg-Fish Phoenix Tail family. In the past it was not recognized by the goldfish industry, it was mostly seen as substandard, that is no longer the case. The picture shows a juvenile Tiger Banded Egg-Fish Phoenix Tail, but its caudal fin has already grown to a mature level resembling that of an adult. Its black stripes are rather pale, it would be better if the fins had more well-defined stripes.

五花蛋凤

Calico Egg-Fish Phoenix Tail (Bluish Base)

- **花色类别** / Color

  软鳞五花 / Calico (Bluish Base)

- **品种类别** / Species

  蛋形 > 原型类 > 蛋凤型

  Egg-Fish > Original Type > Egg-Fish Phoenix Tail

- **品种详述** / Species Description

　　五花蛋凤是传统的名贵品种金鱼，是蛋凤中出现较早的一个花色品种。图例所示为五花蛋凤的亚成体，其作为亚成体来说，色彩和体形发育得都比较好，是较有发展潜质的一尾五花蛋凤。因为五花色的表现，要求体态丰满，颜色才能展现出层次感，因此，五花蛋凤的饲养可以较其他花色的蛋凤适当放宽一些。优秀的五花蛋凤花色最好为红顶白腹素蓝背配芝麻花，其次为素蓝花。老子在《道德经》中说"五色令人目盲"，《庄子•天地》中也说过"五色乱目"，意思都是讲色彩太多太花哨，都易使人眼花缭乱，眩惑难辨。所以，五花色取素雅者为上，在选育中，应避免身上有大块的黑色、红色或黄色斑块。

Calico Egg-Fish Phoenix Tail (Bluish Base) is a rare traditional varieties of goldfish and its occurrence is relatively earlier among Egg-Fish Phoenix Tail family. The picture shows a juvenile Calico Egg-Fish Phoenix Tail (Bluish Base). As a juvenile fish, its color and physique grow good and has great potential for development. Because its appearance performance requires a big and fat body to show its layers, people can feed them a little bit more feeding stuff. Excellent Calico Egg-Fish Phoenix Tail (Bluish Base) has red head, white belly, plain blue with sesame flower, followed by blue pigment. Lao Tzu in *Tao Te Ching* said, "five colors make eyes blind." There is also a record "Chuang-tzu heaven and earth" that, "five colors make eyes confused." It means that too many colors is easy to make people dazzled and confused the enemy. So plain color among Calico is the best. Do not select the Calico Egg-Fish Phoenix Tail (Bluish Base) with big black stain, red stain or yellow stain on the body.

蛋形
Egg-Fish

∨

原型类
Original Type

∨

南京型
Japanese Egg Fish

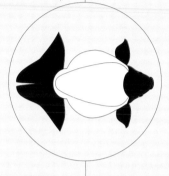

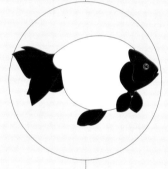

南京是日本金鱼的一个古老代表品种，是由中国的蛋鱼传到日本后逐步培育而成的。南京这个名称是蛋鱼在南方方言中的发音，传到日本后音译而成，与中国城市的地名没有关系。南京在日本有几百年的传承历史，与土佐金、地金以及大阪兰寿并称为日本四大地产金鱼，在日本都是属于古老的地方代表品种。南京的玩赏在日本也属于比较小众，只在小范围爱好者内进行交流。大多数日本红白系金鱼的欣赏，以称之为猩猩的纯红色或红色多于白色的赤胜更纱为贵，而唯独南京是以白色或白色多于红色的白胜更纱为贵。南京的鳞片质地非常特别，具有一种如镜面折射般的七彩光泽。南京的体形比一般的蛋鱼更加圆润丰满一些，如果用人来做比拟，它绝对符合中国盛唐时期对美女的定义标准。南京与日本地金一样，是可以用人工手法对金鱼的体色表现加以干预，体现做鱼乐趣的金鱼品种之一。在日本，南京爱好者们会用梅子发酵获得的一种天然酸性物质梅酸，来涂抹南京的体表，使之红色的部分逐渐褪色为白色，产生自己所需要的花纹颜色。虽然国内对这种做法的认同度不高，但这是源于各民族文化的差异，不可轻率褒贬。南京的体质较弱，对水质变化非常敏感，抗病力也比较差，饲养繁育较困难。近些年来，国内有个别金鱼饲养场和金鱼爱好者引进培育或繁殖，获得成功的为数不多。

Nankin is an ancient representative breed of Japanese goldfish varieties which was cultivated from Egg-Fish sent to Japan from china. The name "Nankin" is the southern dialect pronunciation of Egg-Fish and is transliteration after being sent to Japan. It has nothing to do with the name of city in china. (In china, there is a city named Nanjing). Nankin has hundreds of years' history in Japan, together with Tosakin, Jikin, and Osaka Ranchu known as the four largest domestic goldfish. They are ancient local representative varieties. In Japan, Nankin enjoys a relatively small minority and only be communicated within a small range of enthusiasts. As for the appreciation of most Japanese Red White gold fish, the Shojo red or Akagachisarasa with more red is precious. But for Nankin, the white or the Shirogachisarasa with more white is precious. Its scale texture is very special with shiny colorful luster from the mirror reflection. Compare with the normal Egg-Fish, Nankin is more bigger and fatter. If consider Nankin as human being, it is absolutely conforms to the standard definition of the beauty of Tang Dynasty in China (in Tang Dynasty, as for females, more fullness is more beautiful.) Like Jikin, Nankin is one of varieties of which the colors could be controlled by artificial intervention for the joy of being fish. In Japan, people who love Nankin apply natural acid substance plum acid by plum fermentation on its surface in order to make red color fade away into white, and then to get the color they want. Although the differences of national culture make the domestic acceptance of such practices not high, we can not judge this practice is good or not. It is very hard to breed, because its physique is not so strong, it is very sensitive about the changes in water quality, and its disease resistance is relatively poor. In recent years, individual domestic goldfish farms and some goldfish enthusiasts have introduced it to breed, however few of them are succeed.

花色类别 / Color •

红白 / Red White

品种类别 / Species •

蛋形 ＞ 原型类 ＞ 南京型
Egg-Fish ＞ Original Type ＞ Japanese Egg-Fish

品种详述 / Species Description •

出云南京
Izumo Nank n (Japanese White Egg-Fish wih Six Red Patches)

　　南京是日本古老金鱼的代表品种之一，因大多出产自日本岛根县的出云地区，因此，大家又习惯称之为出云南京。岛根县的出云地区，大致对应日本古代传说中的出云国。在日本神话传说中，这里是居住着许多神灵的国度，如同中国神话传说中的蓬莱、方丈和瀛洲一样，给人们以美好的遐想。出云二字在汉语中可以引申为超出云彩的含义，恰巧南京又是以白色为贵，因此，以出云南京命名已不仅局限于表达地区代表鱼种，更是日本人对传统文化与金鱼文化更具深远、浑然天成、最妙不过的融合与用意。

Nankin is one of ancient representative breeds of Japanese goldfish varieties. Because most of them breed in Shimane Prefecture of Japan Izumo region, it is used to be called Izumo Nankin (Japanese White Egg-Fish with Six Red Patches). Shimane Prefecture Izumo region, is roughly Japan Izumo Province, in Japanese mythology, it is inhabited by many gods, as Chinese myths and legends of the Penglai, the abbot and Yingzhou, to give people a beautiful reverie. "Out cloud" in Chinese means beyond the cloud. Coincidentally, as for Nankin, the white is the best, therefore, the name Izumo Nankin (Japanese White Egg-Fish with Six Red Patches) is not only a local representative variety but also the Japanese profound and wonderful intention to naturally fusion between traditional culture and goldfish culture.

蛋形
Egg-Fish

∨

头型变异类
Head Variation

∨

鹅头型
Goosehead

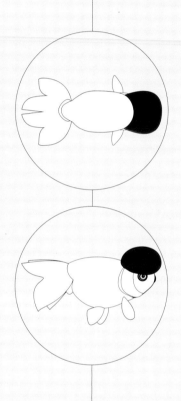

金鱼的头型变异中，不论文形金鱼还是蛋形金鱼，均可划分为鹅头型、虎头型、狮头型和龙头型四大类。按进化论中从低级到高级、从简单到复杂的原则，鹅头型应该是出现比较早的头型变异类型，因其头形酷似鹅的头茸，故而得名。有些地方又将鹅头型叫做高头型，其主要特点是顶茸比较发达，而面颊部及鳃盖部均相对比较光滑，无赘生物。在鹅头型金鱼的鉴赏中，以顶茸发达，质地细密，出檐而不遮目者为贵。鹅头型是头型变异类较早期的一个品种类型，目前保留的不多，蛋形金鱼中，鹅头型最具代表性的就是北京的鹅头红、凤尾鹅头红、五花鹅头以及目前金鱼爱好者杂交选育的蓝鹅头和紫鹅头等。20世纪80年代还可以看到五花鹅头，目前可能已经灭绝。近些年来，有金鱼爱好者在鹅头红的基础上，又发展出复合变异的龙睛鹅头红，亦比较有趣，如能稳定，又为中国金鱼添加了一个新的品种。

Among Head Variation of gold fish, no matter Fantail Goldfish or Egg-Fish, they can be divided into 4 categories: Goosehead, Tigerhead, Lionhead and Dragonhead. According to the principle of evolution theory, from low level to high level, from simple to complex, Goosehead should be the relatively earlier Head Variation type. Because its head shape so resembles the pompons on goose head, it earns the name. In some place, Goosehead is called High-Head. Its main features are as follows: head pompon is developed; cheek and operculum department are relatively smooth without neoplasm. Goosehead with developed head pompon, fine-grained, the canopy non-covering eyes is the best. Goosehead is a early type of head varieties. The reservation is not so much. Among Egg-Fish the most representative Goosehead types are Beijing Red Cap Goosehead, Red Cap Goosehead Phoenix Tail, Calico Goosehead (Bluish Base), Blue Goosehead Phoenix Tail and Chocolate Goosehead Phoenix Tail. Calico Goosehead (Bluish Base) existed during the 20th century 80 years, at present has probably wiped out. During recent years, on the basis of Red Cap Goosehead, some goldfish breeders have developed Red Cap Goosehead with Dragon eyes of compound variation which is also very interesting. If it can be stable raised, it will add a new variety for Chinese goldfish.

鹅头红 / Red Cap Goosehead

花色类别 / Color •

红白 / Red White

品种类别 / Species •

蛋形 > 头型变异类 > 鹅头型
Egg-Fish > Head Variation > Goosehead

品种详述 / Species Description •

　　鹅头红是蛋形金鱼中，头型变异类鹅头型的代表品种，也是北京金鱼最著名的代表品种之一，已故的著名学者、金鱼鉴赏大师刘景春先生对鹅头红有很高的评价。由于多种原因，鹅头红曾数度濒临灭绝。1984年，鹅头红曾作为中国金鱼的代表品种之一，远赴美国纽约展览并引起轰动。当时，李振德先生将它称为宫廷金鱼，因此，鹅头红又称之为宫廷鹅头红。鹅头红身姿秀美，虽是北派金鱼代表，却充满南方灵秀之气。其色彩搭配是最让人喜爱的红白组合，其全身洁白如玉，唯有头顶嵌有一颗红宝石顶冠，十分华贵。鹅头红体质比较弱，对水质和饵料要求较高，是比较难饲养的金鱼品种之一。

Red Cap Goosehead is representative breed of head variation varieties in Egg-Fish family. It is also one of the most famous representative breed of Beijing goldfish. The late famous scholar, Mr. Liu Jingchun, a master of goldfish appreciation, has a very high evaluation of the Red Cap Goosehead. Due to a variety of reasons, Red Cap Goosehead had been on the verge of extinction for several times. In the year of 1984, Red Cap Goosehead as one of representative varieties of Chinese gold fish had been sent to the United States for New York exhibition which caused a sensation. 1984, goose red had as one of the representative varieties of goldfish Chinese, went to the United States and New York exhibition caused a sensation. At that time Mr. Li Zhende called it Palace goldfish, therefore, Red Cap Goosehead is also named Palace Red Cap Goosehead. It has a beautiful figure. Although it is the representative of the northern school, it is full of southern aura. Its color matching is the most delightful red and white. It has totally white body except for a ruby crown on its head, very luxurious. Red Cap Goosehead physique is weak so it need high quality of water and is one of the difficult keeping goldfish varieties.

凤尾鹅头红

Red Cap Goosehead
Phoenix Tail

- **花色类别** / Color
  红白 / Red White

- **品种类别** / Species
  蛋形 > 头型变异类 > 鹅头型
  Egg-Fish > Head Variation > Goosehead

- **品种详述** / Species Description

　　凤尾鹅头红又称凤鹅或鹅头丹凤，是鹅头红中的一个长尾变异品种，后被逐步稳定固化。和鹅头红一样，凤尾鹅头红的鳞片要求质地细密，有冷银色光泽，排列整齐。其顶茸发育适中，颜色鲜艳。而相较鹅头红而言，凤尾鹅头红身形略长些较好，头身比在1：2.5～1：3为佳，同蛋凤类似，体态贵瘦不贵腴。所有的鱼鳍也相对变长，尤其是二龄以后，其狭长的尾鳍，如同凤凰的尾羽飘带一般，更有一份绰约的仙姿。目前也有利用杂交选育的方式培养出来的凤尾鹅头红，其各自的风格也有所差异。

Red Cap Goosehead Phoenix Tail, also named Phoenix-Goose, is a long tail variation of Red Cap Goosehead family and late is gradually stable and fixed. Just like Red Cap Goosehead, Red Cap Goosehead Phoenix Tail the scale texture is fine with cold silvery color, and neatly arranged. Compared with Red Cap Goosehead, it is better for Red Cap Goosehead Phoenix Tail to have a longer body. The proportion of head and body is better between 1:2.5-1:3. It is similar as Egg-Fish Phoenix Tail, thin is better than fat. All the fins grow longer especially after 2 years old, its caudal fin is narrow and long, just like the tail feather ribbon with graceful beauty. At present there is also Red Cap Goosehead Phoenix Tail by hybrid breeding cultivation. Their respective styles are so vary.

花色类别／Color •

蓝／Blue

品种类别／Species •

蛋形 > 头型变异类 > 鹅头型

Egg-Fish > Head Variation > Goosehead

品种详述／Species Description •

凤尾蓝鹅头 — Blue Goosehead Phoenix Tail

　　图例所示为凤尾蓝鹅头，是近些年来金鱼业者通过杂交选育的方式培育出来的。其体形比鹅头红略显雄健，顶茸发达可出檐，质地细腻，属几较优为的再形蛋由鹅头型个体。其外形通体为蓝黑色并带有金属质感亮银色反光，鱼鳍较长，质厚。由于选育时间不长，其遗传稳定性较弱，如果假以时日定向培育，一定可以逐步稳定，成为一个新的优秀金鱼品种。目前凤尾蓝鹅头存世量稀少，较珍贵。

As shown in the illustration is Blue Goosehead Phoenix Tail. It is cultivated through hybrid breeding by goldfish industry. Its size is slightly more vigorous than Red Cap Goosehead. Head pompon is developed and canopy, fine texture. It is a good Goosehead individual of Egg-Fish. It is totally blue and black with metallic bright silver reflection. Its fins are relatively long and thick. Because the cultivation time is not so long, the breeding is not so stable. If cultivate it for a long time, it must be stable finally and become a new excellent goldfish variety. Currently, the amount of Blue Goosehead Phoenix Tail is few in the world so it is relatively rare.

蛋形
Egg-Fish
∨
头型变异类
Head Variation
∨
虎头型
Tigerhead

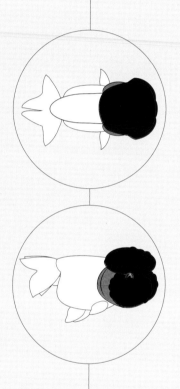

虎头型大多数个体在顶茸上，有一条阴刻的纵线和三四道阴刻的横线，颇似老虎前额的花纹，故而得名。虎头型是介于鹅头型和狮头型的一个中间类型，它的特点是：除具有和鹅头型一样较发达的顶茸外，面颊部以及目下均有发达的鬓茸，有吻凸但并不夸张，鳃盖上无明显鳃茸，较干净且平滑。在蛋形金鱼中，头型变异类为虎头型的金鱼并不多，主要代表品种为北京的红虎头，日本金鱼秋锦也可以归入这个变异类型中。目前在红虎头的基础上，金鱼业者也逐步培育出各种花色的虎头型，如红白虎头、蓝虎头、紫虎头、黑虎头等等。

Most Tigerhead individuals have a vertical incised lines and three or four incised horizontal lines on the head pompon resembling the patterns on tigers forehead, so it earns its name. Tigerhead is an intermediate type between Goosehead and Lionhead, which is characterized by: in addition to having the same developed head as Goosehead, the cheek are developed as sideburns pompon, the snout is not exaggerated and the operculum is without obvious sideburns pompon, relatively clean and smooth. Tigerhead variation among Egg-Fish family is not so much. The main representative variety is Red Lionhead in Beijing. Japanese goldfish Shukin is also this variation. At present, based on the Red Lionhead, Tigerhead in various colors has been gradually cultivated by goldfish industry such as Red White Lionhead (Tigerhead Type), Blue Lionhead (Tigerhead Type), Chocolate Lionhead (Tigerhead Type), Black Lionhead (Tigerhead Type) and so on.

红虎头 — Red Lionhead (Tigerhead Type)

花色类别 / Color •

红 / Red

品种类别 / Species •

蛋形 > 头型变异类 > 虎头型

Egg-Fish > Head Variation > Tigerhead

品种详述 / Species Description •

　　红虎头又称王虎、王字虎头，是蛋形金鱼头型变异类中虎头型的代表品种，它与鹅头红并列作为北京金鱼的代表品种。已故著名学者、金鱼鉴赏大师刘景春先生对红虎头有很高的评价，称之为有压倒一切之优势的品种，虽有些夸张，却足见老先生对其喜爱至深。1960年6月1日，我国发行的特38《金鱼》邮票，其中第4枚便是红虎头。刘景春先生对它概括性的描述是：红虎头背润且阔、头高且大、唇凸嘴窝、见棱见角。老北京饲养鹅头红和红虎头常用大木海或陶制的虎头盆，如同慢火煲靓汤，当岁墩养，二年以后才逐步放开身形，三龄以后进入最佳观赏阶段。

Red Lionhead also named Wang Lionhead (Tigerhead Type) (the tiger with the Chinese character 'Wang' pattern on its forehead), is the representative breed of Tigerhead in Egg-Fish family. Both Red Lionhead and Red Cap Goosehead are representative breeds of Beijing goldfish. The late famous scholar, Mr. Liu Jingchun, a master of goldfish appreciation, has a very high evaluation of the Red Lionhead. He said, it is over all the advantages of other varieties. Although some exaggeration, well can know that this old gentlemen is in deep love with Red Lionhead. In special "goldfish" stamps , issued on June, the 1st , 1960, the fourth is Red Lionhead. Mr. Liu Jingchun has given a general description of it, the back is moist and wide, the head is high and large, the lips are convex, the mouth is nest with angles and corners. In old Beijing, people usually keep Red Cap Goosehead and Red Lionhead in a pot with Tigerhead made up of wooden (Da mu hai in chinese) or ceramics, as simmered soup. It begins to grow big after 2 years. When it is 3 years old, it is the best time to view.

长尾蓝虎头

Blue Lionhead (Tigerhead Type) Long Tail

- **花色类别** / Color

  蓝 / Blue

- **品种类别** / Species

  蛋形 ＞ 头型变异类 ＞ 虎头型

  Egg-Fish ＞ Head Variation ＞ Tigerhead

- **品种详述** / Species Description

  图例所示为长尾蓝虎头，是近些年来杂交选育的新品种，外部形态和日本金鱼秋锦类似，其体态雄壮，长尾；全身蓝色，有部分个体颜色不稳定，会逐渐变为蓝白色，最后转白。而秋锦由于遗传稳定，其转色的比例较少。长尾蓝虎头由于选育的时间较短，遗传稳定性较差，因此还不能完全确立为一个新的金鱼品种。

  As shown in the illustration is Blue Lionhead (Tigerhead Type) Long Tail is a new breed hybridized cultivated recently. Its outlook is similar as Shukin, with a strong physique and a long tail. it is totally blue except that some individuals color is not stable and will gradually turn into blue and white, finally white. While, due to Shukin has stable heredity, its color turning proportion is less. The time of cultivating Blue Lionhead (Tigerhead Type) Long Tail is short and its heredity is unstable so it cannot exist as a new goldfish breed.

花色类别 / Color •

紫 / Chocolate

品种类别 / Species •

蛋形 > 头型变异类 > 虎头型

Egg-Fish > Head Variation > Tigerhead

品种详述 / Species Description •

紫虎头

Chocolate Lionhead
(Tigerhead Type)

　　图例所示为紫虎头。紫虎头多为蛋形金鱼头型变异类中的狮头型，此尾可能是其中出现的突变个体，也可能是通过杂交选育获得的。其头型特征符合虎头型的标准，体态特征近似红虎头，雄浑敦厚，尾鳍中等，全身青紫色，有紫色王虎的意趣。由于比较罕见，因此较为珍稀。可以选择优秀个体定向选育，使其逐步稳定，成为一个优秀的金鱼新品种。

As shown in the illustration is Chocolate Lionhead (Tigerhead Type). Chocolate Lionhead (Tigerhead Type) is most probably lion head in goldfish head variation. Maybe its tail is just a variation individual or from hybrid cultivation. Its head feature meet the standards of Tigerhead; physic feature is similar as Red Lionhead, strong and thick, caudal fin is not very long or not very short; it is totally cyanosed, with Chocolate Wang Lionhead (Tigerhead Type) charm. It is very rare to see it. People can targeted cultivate some excellent individuals, making their heredity stable, to be a good new breed of goldfish.

紫 / Chocolate

蛋形
Egg-Fish

∨

头型变异类
Head Variation

∨

狮头型
Lionhead

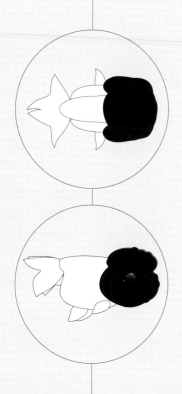

在蛋形金鱼头型变异类中，狮头型是品种数量最大的一个类别，也是最受大家喜爱的一个类别。由于地域文化差异以及流传时的讹误，也有许多地方将蛋形金鱼中的虎头型和狮头型，都称之为"虎头"。狮头型的特点在于：整个头部都有发达的头茸分布，包含鳃盖甚至到下颌部亦是如此，如同非洲雄狮包裹着威武的狮鬃一般，故而得名。由于受到地域饲养手法的差异，以及审美风格追求差异的影响，目前狮头型主要有四大风格类型：虎头类、猫狮类、寿星类和兰寿类。四大类型虽然外形风格迥异，但头茸分布均符合狮头型的定义标准。虎头、猫狮和寿星均为中等体形，大多以直背者为贵，观赏多取俯视。而兰寿一般多为梳子背，体形偏大，观赏则以侧视为佳。

Among head variation variations of Egg-Fish, Lionhead contains the largest number of species, and is also the most popular. In many places both Tigerhead and Lionhead in Egg-Fish family are called "Lionhead" because of the local culture differences and spread corruption. It is characterized in that: the whole head if full of developed head pompon which even cover the part from the gill cover to the lower jaw, as the African lion wraps mighty lion's mane, so it earns the name Lionhead. Due to the regional differences in breeding practices and the impact of pursuit of aesthetic style, at present, Lionhead has mainly four types: Tigerhead, Lionhead (with Well Developed Head Growth and Chubby Body), Lionhead (Southern) and Ranchu. Although the four types are in different shapes, all their well-distributed head pompon are in accordance with the standard definition of Lionhead. Lionhead (Tigerhead Type), Lionhead (with Well Developed Head Growth and Chubby Body) and Lionhead (Southern) are medium-sized, and mostly the straight back is the best, waist-level view. However, in general, Ranchu mostly has comb back, bigger body which is suitable for side observation.

花色类别 / Color •

黑 / Black

品种类别 / Species •

蛋形 ＞ 头型变异类 ＞ 狮头型
Egg-Fish ＞ Head Variation ＞ Lionhead

品种详述 / Species Description •

　　图例所示为黑猫狮。猫狮为武汉金鱼的代表品种，以前称为狮子头，又有方头和圆头的区分。猫狮与其他几种蛋形金鱼狮头型的区分在于头部发育偏横向取势，肉茸及吻凸发达，颗粒较大，游动时头茸偶有颤动，十分有趣。猫狮的体形相对其他几类有些偏小，所以命名猫狮有狮子头小猫身的含义。武汉当地饲养猫狮喜用水蚯蚓，其高蛋白对头茸的发育有决定性的作用。

As shown in the illustration is Black Lionhead (with Well Developed Head Growth and Chubby Body) is representative breed of Wuhan goldfish, called Lionhead in the past, divided into square head and round head. The differences between Lionhead (with Well Developed Head Growth and Chubby Body) and other Lionhead in Egg-Fish family are that: the head grows in horizontal trend, the whole head pompon and snout are well-developed, larger particles, and the head pompon quivers occasionally when it is swimming which is very interesting. The body of the Lionhead (with Well Developed Head Growth and Chubby Body) is somewhat smaller than the others, so it has the name of Lionhead (with Well Developed Head Growth and Chubby Body) which means the head as a lion and body as a cat. In Wuhan, people feed Lionhead (with Well Developed Head Growth and Chubby Body) with water earthworm full of high protein which plays a decisive role in the growth of the head pompon.

黑猫狮 — Black Lionhead (with Well Developed Head Growth and Chubby Body)

黑 / Black

蓝猫狮

Blue Lionhead (with Well Developed Head Growth and Chubby Body)

- **花色类别** / Color
  蓝 / Blue

- **品种类别** / Species
  蛋形 > 头型变异类 > 狮头型
  Egg-Fish > Head Variation > Lionhead

- **品种详述** / Species Description

  蓝猫狮头呈倒八字形，多细茸，横向取势，其鬓茸异常发达；身体较武汉猫狮粗壮，身形偏短，活跃度高。通过几年的定向选育，其遗传较为稳定，成为唐山地区较为突出的地方代表品种，并多次在国内金鱼大赛中获奖。

  The head of Blue Lionhead (with Well Developed Head Growth and Chubby Body) appears to be a upside-down 8, with a lot of exquisite pompon growing horizontally and unusually developed sideburns pompon. It is sturdier but shorter than Wuhan Lionhead (with Well Developed Head Growth and Chubby Body), and very active. After several years' directive selection and breeding, its inheritance has become relatively stable, and then it becomes an outstanding local representative variety of Tangshan and has won several prizes in the domestic goldfish competitions.

红白猫狮 ｜ Red White Lionhead (with Well Developed Head Growth and Chubby Body)

花色类别 ／ Color •

红白 ／ Red White

品种类别 ／ Species •

蛋形 ＞ 头型变异类 ＞ 狮头型
Egg-Fish ＞ Head Variation ＞ Lionhead

品种详述 ／ Species Description •

　　图例所示为红白猫狮，是近些年来武汉金鱼业者，从大红猫狮中选育出的一个新的猫狮花色品种。红白猫狮头呈倒八字形，粗茸大颗粒，横向取势，其鳃茸、鬓茸异常发达，吻凸发育较好。其身形相对唐山五花猫狮较狭长，与发达的头部形成明显对比，并与之平衡。红白猫狮中最难得的是与红顶虎头极为相似的红顶白猫，武汉又称之为猫鹅，十分稀少。

The legend shows Red White Lionhead (with Well Developed Head Growth and Chubby Body), which is a new Lionhead (with Well Developed Head Growth and Chubby Body) variety selected and bred from Bright Red Lionhead (with Well Developed Head Growth and Chubby Body) by the goldfish farmers in Wuhan in recent years. The head of Red White Lionhead (with Well Developed Head Growth and Chubby Body) appears to be upside-down 8, with large-particle and thick pompon growing horizontally, unusually developed gill pompon and sideburns pompon, and the snout is well-developed. Compared to Tangshan Calico Lionhead (with Well Developed Head Growth and Chubby Body) (Bluish Base), its body is narrower and longer, presenting a striking contrast with the developed head and keeping a balance with it. Among Red White Lionhead (with Well Developed Head Growth and Chubby Body), the rarest one is Red Cap White Lionhead (with Well Developed Head Growth and Chubby Body), which is very similar to Red Cap Lionhead, and in Wuhan, it is called Cat-Goose, and is extremely rare.

红黑猫狮
—— Red Black Lionhead (with Well Developed Head Growth and Chubby Body)

- **花色类别** / Color

  红黑 / Red Black

- **品种类别** / Species

  蛋形 > 头型变异类 > 狮头型

  Egg-Fish > Head Variation > Lionhead

- **品种详述** / Species Description

  图例所示为红黑猫狮，头茸发育横向取势，两鬓和鳃盖均有发达的头茸，吻凸亦很发达，前后融为一体，形成标准的方头。这种方头型的猫狮曾在国内各类金鱼大赛中取得优异的赛绩，也广受猫狮粉们的喜爱。红黑色是向红色转色的一个过渡色，褪色完成后，其通体呈现出如夏天傍晚火烧云一般的艳美朱砂红色。猫狮头茸十分发达，但对其平衡能力要求严格，其游动起来要雄健霸气，如果趴缸，其欣赏价值就会大打折扣。

As shown in the illustration is Red Black Lionhead (with Well Developed Head Growth and Chubby Body). Its head pompon grows in horizontal trend. Both sideburns and gill covers have developed head pompon and snout is also well developed. They are integrated from front and rear to form a standard square head. This type of Lionhead (with Well Developed Head Growth and Chubby Body) with square head have achieved excellent scores in various types of domestic goldfish contest. It is widely popular with its enthusiasts. Red and black is a transition color before it turns into red. After the completion of the black-fading, the whole body is as beautiful as the summer evening glow. Its head velvet is well developed but the requirement of its ability to balance is very strict. When swimming , it is supposed to be in vigorous ambition. If lying in the cylinder, its appreciation value will be greatly reduced.

撒
锦
猫
狮

—— Lionhead (with Well Developed Head Growth and Chubby Body)(with Bluish Matt and Black Dots)

- **花色类别** / Color

  软鳞五花 / Bluish Matt and Black Dots

- **品种类别** / Species

  蛋形 > 头型变异类 > 狮头型

  Egg-Fish > Head Variation > Lionhead

- **品种详述** / Species Description

  　　撒锦猫狮是从五花猫狮中选育出的一个猫狮花色品种。撒锦猫狮外形和五花猫狮没有太大差异，蓝白的底色点缀少许细碎黑点，花色较为素雅。图例所示的撒锦猫狮身形适中，头茸颗粒与武汉猫狮较为接近。饲养撒锦猫狮和五花猫狮时要注意，夏秋适度遮光对其色彩的稳定有较关键的功效。

  Lionhead (with Well Developed Head Growth and Chubby Body) (with Bluish Matt and Black Dots) is a Lionhead (with Well Developed Head Growth and Chubby Body) variety selected and bred from Calico Lionhead (with Well Developed Head Growth and Chubby Body) (Bluish Base). There is not great difference between Lionhead (with Well Developed Head Growth and Chubby Body) (with Bluish Matt and Black Dots) and Calico Lionhead (with Well Developed Head Growth and Chubby Body) (Bluish Base) in appearance. The bluish base is dotted with few black small breaking points and the colors are quiet. The Lionhead (with Well Developed Head Growth and Chubby Body) (with Bluish Matt and Black Dots) in the legend is moderate in body shape, and its head pompon particles are similar to those of Wuhan Lionhead (with Well Developed Head Growth and Chubby Body). When feeding Lionhead (with Well Developed Head Growth and Chubby Body) (with Bluish Matt and Black Dots) and Calico Lionhead (with Well Developed Head Growth and Chubby Body) (Bluish Base), the feeders should realize that the proper shading in summer and autumn has a critical effect on the stability of colors.

麒麟猫狮 | Kirin Lionhead (with Well Developed Head Growth and Chubby Body)

- **花色类别** / Color
  软鳞五花 / Kirin

- **品种类别** / Species
  蛋形 > 头型变异类 > 狮头型
  Egg-Fish > Head Variation > Lionhead

- **品种详述** / Species Description

　　麒麟猫狮是武汉金鱼业者，通过五花虎头（狮头型）杂交选育出的一个猫狮花色品种。图例所示麒麟猫狮外形和麒麟虎头（狮头型）比较接近，头茸方正，颗粒适中，鬃茸和鳃茸发育适度，不及传统的大红猫狮发达；鳞片基部黑灰色，边缘为蓝色或反光的亮银色，全身呈青黑色，杂色少，属于墨麒麟花色。麒麟猫狮的头茸发育度还应继续培育，使之更具猫狮的风采。

Kirin Lionhead (with Well Developed Head Growth and Chubby Body) is a Lionhead (with Well Developed Head Growth and Chubby Body) variety selected and bred through hybridization of Calico Lionhead (Bluish Base) (Lionhead Type) by the goldfish farmers in Wuhan. In appearance, the Kirin Lionhead (with Well Developed Head Growth and Chubby Body) in the legend is similar to Kirin Lionhead (Lionhead Type), with upright head pompon, moderate particles, and moderately-developed gill pompon and sideburns pompon, which are less developed than those of the traditional Bright Red Lionhead (with Well Developed Head Growth and Chubby Body). The base of the scales is black gray while the edges are blue or reflective sprightly silver, and the whole body is blue black with few other colors, belonging to Black Kirin color. The head pompon of Kirin Lionhead (with Well Developed Head Growth and Chubby Body) should be continually developed so as to be a more representative Lionhead (with Well Developed Head Growth and Chubby Body).

花色类别 / Color •

软鳞五花 / Calico(Bluish Base)

品种类别 / Species •

蛋形 > 头型变异类 > 狮头型

Egg-Fish > Head Variation > Lionhead

品种详述 / Species Description •

五花猫狮 —— Calico Lionhead (with Well Developed Head Growth and Chubby Body) (Bluish Base)

图例所示为五花猫狮，是近些年来，由河北唐山金鱼业者，通过武汉红猫狮和五花虎头（狮头型）杂交选育出的一个新的猫狮花色品种。五花猫狮头型类似武汉猫狮中的圆头型，头茸颗粒适中，总体呈横向取势，较浑圆，其鬓茸比较发达。身体较武汉猫狮粗壮，身形偏短，憨态可掬。其头、身、尾的比例近似1：1：1，是否对平衡会有影响，还需一定的时间来检验。其在花色上的培育值得肯定。

The legend shows Calico Lionhead (with Well Developed Head Growth and Chubby Body) (Bluish Base), which is a new Lionhead (with Well Developed Head Growth and Chubby Body) variety selected and bred through hybridization of Wuhan Red Lionhead (with Well Developed Head Growth and Chubby Body) and Calico Lionhead (Bluish Base) (Lionhead Type) by the goldfish farmers in Tangshan, Hebei in recent years. The head of Calico Lionhead (with Well Developed Head Growth and Chubby Body) (Bluish Base) is similar to the round head of Wuhan Lionhead (with Well Developed Head Growth and Chubby Body), with moderate head pompon particles growing horizontally and relatively developed sideburns pompon. It is sturdier but shorter than Wuhan Lionhead (with Well Developed Head Growth and Chubby Body), and looks charmingly naïve. The proportions of its head, body and tail are approximately 1:1:1. And it will take more time to know whether this will affect the balance. And its breeding in color is worth approving.

黑虎头

Black Lionhead (Northern)

- **花色类别** / Color

  黑 / Black

- **品种类别** / Species

  蛋形 > 头型变异类 > 狮头型

  Egg-Fish > Head Variation > Lionhead

- **品种详述** / Species Description

图例所示为黑虎头，属于北方的虎头风格，头茸发育紧实，横向和纵向取势较为均衡，顶茸与两侧头茸有较明显分界线，呈块状分布，俗称三块瓦。黑虎头以全身皆若墨染者为佳，尤其腹部不能泛出青色或黄色。此鱼优点是头部发育较为完美，缺憾在于头身比例约为1∶1.5，略微偏短，背有些弓，可能会制约以后的发展。饲养黑虎头有驱凶辟邪的寓意，因此广受市场青睐。一般红虎头和五花虎头中青色个体都会出现黑虎头，而后者出现的颜色较为稳定。

As shown in the illustration is Black Lionhead (Northern) which is the northern Lionhead style, with tight head pompon, well distributed horizontally and vertically. There is a clear dividing line between head pompon and both sides head pompon, distributed in blocks, commonly known as three watts. Black Lionhead (Northern) with totally black body is the best, especially the belly should not display blue or yellow. Its advantage is that the head developed perfectly and its disadvantage is that the proportion of head and body is about 1;1.5, little bit short, and the back is little bit bow which may interrupt its future development. Keeping Black Lionhead (Northern) means driving fierce and evil spirits so it is widely popular in the market. Generally speaking, there is Black Lionhead (Northern) from all blue individuals in Red Lionhead (Northern) and Calico Lionhead (Bluish Base) (Northern) and that from Calico Lionhead (Bluish Base) (Northern) the color is more stable.

花色类别 / Color •
蓝 / Blue

品种类别 / Species •
蛋形 > 头型变异类 > 狮头型
Egg-Fish > Head Variation > Lionhead

蓝虎头 — Blue Lionhead (Northern)

品种详述 / Species Description •

　　图例所示为蓝虎头，属于北方的虎头风格，头茸发育紧实，横向和纵向取势较为均衡，圆头形，形状及质感都酷似淮扬菜系中的名菜狮子头。蓝虎头的色泽根据遗传基因和水质情况分为青蓝色和墨蓝色两种。青蓝色偏银亮，墨蓝色偏黑，图例为青蓝色的虎头。同所有的虎头一样，蓝虎头也应以背直且平者为贵，弯背者次之。受温度及水质等因素的影响，蓝虎头中会有个体出现褪色为黑白或喜鹊花以至于白色的现象。

As shown in the illustration is Blue Lionhead (Northern), which is northern Lionhead style. Its head pompon is tight, well-distributed horizontally and vertically. It has round head of which the shape and texture resemble famous dishes "Lionhead" in Huaiyang cuisine. According to the genetic gene and water quality, the color luster of Blue Lionhead (Northern) can be divided into two, cyan-blue and dark-blue. Cyan-blue is more silver and dark-blue is more black. As shown in the illustration is cyan-blue Lionhead (Northern). Like all Lionhead, Blue Lionhead (Northern) with straight and flat back is the best, followed by what the back is bow. Due to the factors of temperature and water quality, some Blue Lionhead individuals will fade into black and white, or magpie flowers resulting white.

紫虎头 | Chocolate Lionhead (Northern)

- **花色类别** / Color
  紫 / Chocolate

- **品种类别** / Species
  蛋形 > 头型变异类 > 狮头型
  Egg-Fish > Head Variation > Lionhead

- **品种详述** / Species Description

　　图例所示为紫虎头。紫虎头在过去图谱中未曾有过记载，是利用雪青虎头杂交选育出的一个新品种，后又流传至北京、山东等地，当地金鱼业者又有进一步的改良和提纯。紫虎头由于是杂交选育的品种，体形较大。其头茸受雪青虎头的影响，发育较晚，约二龄以后才发育显现，三龄以后达到最佳观赏期，属于晚成型金鱼品种。

As shown in the illustration is Chocolate Lionhead (Northern). There is no record about Chocolate Lionhead (Northern) in the old collection of illustration. It is a new breed hybrid cultivated from Lilac Lionhead (Northern), and then spread to Beijing and Shandong area, where the local goldfish industry has further improved and purified. Because it is hybrid cultivated, it is in big size. Its head pompon develops late because of the impact of Lilac Lionhead (Northern). Up till after two years stage, it will obviously develop. It is the best period to view it after it is in 3 years stage. It is a goldfish variety that develops slowly.

花色类别／Color ●

雪青／Lilac

品种类别／Species ●

蛋形 ＞ 头型变异类 ＞ 狮头型

Egg-Fish ＞ Head Variation ＞ Lionhead

品种详述／Species Description ●

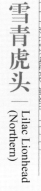

雪青虎头 ——Lilac Lionhead (Northern)

　　图例所示为雪青虎头，属于北方的虎头风格，头茸发育紧实，横向和纵向取势较为均衡，顶茸与两侧头茸有较明显分界线，呈块状分布，俗称三块瓦。以往雪青虎头一直有层神秘的面纱，未曾出现在世人眼中，过去的图谱中也未曾记载过，其出现的年代应该较晚。雪青虎头为京津地区的金鱼业者通过蓝虎头选育而出的。其存世量稀少，正品率较低，属于珍稀品种。雪青虎头头茸发育较晚，约二龄以后才发育显现，三龄以后达到最佳观赏期，属于晚成型金鱼品种。

As shown in the illustration is Black Lionhead (Northern) which is the northern lionhead style, with tight head pompon, well distributed horizontally and vertically. There is a clear dividing line between head pompon and both sides pompon, distributed in blocks, commonly known as three watts. In the past Lilac Lionhead (Northern) was always covered with a mysterious veil, having never appeared in the eyes of the world. There is either no record about Lilac Lionhead (Northern) in the old collection of illustration. The occurrence of the era should be late. It is cultivated from Blue Lionhead (Northern) by goldfish industry staff in Beijing and Tianjin area. The amount of it is very small in the world and the genuine rate is low, so it is rare breed. Its head pompon develops late. Up till after two years stage, it will obviously develop. It is the best period to view it after it is in 3 years stage. It is a goldfish variety that develops slowly.

雪青／Lilac

红白虎头 — Red White Lionhead (Northern)

- **花色类别** / Color
  红白 / Red White

- **品种类别** / Species
  蛋形 > 头型变异类 > 狮头型
  Egg-Fish > Head Variation > Lionhead

- **品种详述** / Species Description

　　图例所示为红白虎头，属于南方的虎头风格，头茸发达但颗粒大而松，横向和纵向取势较为均衡，顶茸与两侧头茸分界线不太明显，方头型与圆头型兼有。红白虎头发育较早，当岁鱼即可起头。但饲养时若不加以适当控制，红白虎头二龄后头茸会散，影响其观赏价值。因此，红白虎头的饲养应该借鉴北方传统金鱼的饲养手法，适当控制其生长节奏，以达到最佳的观赏效果。红白虎头性格较活泼，易与人亲近，多为俯视欣赏。

As shown in the illustration is Red White Lionhead (Northern) which is the southern Lionhead style with well-developed head pompon but in big particles and loose, which is well distributed horizontally and vertically. The dividing line between head pompon and both sides pompon is not so clear. It has square head and round head types. It develops early, when it is in 1 year stage the head has began growing. However, if not properly controlled when feeding, its head pompon will be scattered after 2 year stage which will affect its ornamental value. Therefore, Red White Lionhead (Northern) breeding should refer to northern traditional goldfish breeding practices. Properly control its growth rhythm to achieve the best ornamental effect. Red White Lionhead (Northern) is lively and easy to close. It is mostly suitable for waist-level viewing.

红顶虎头

Red Cap Lionhead (Northern)

- **花色类别** / Color

  红白 / Red White

- **品种类别** / Species

  蛋形 > 头型变异类 > 狮头型

  Egg-Fish > Head Variation > Lionhead

- **品种详述** / Species Description

　　红顶虎头俗称红顶虎，武汉地区又称之为猫鹅，是狮头型虎头中最令人瞩目的明星。江苏无锡的许祺源先生及苏州的老金鱼把式马宝根先生等金鱼前辈，通过多年的定向培育，将红白虎头中的红顶个体与猫狮杂交选育，并不断提纯从而获得红顶虎头。其头型特征上偏向于武汉猫狮，横向取势，两鬓及鳃茸发达，头顶正中嵌入一块方形的红色印记。其身形适中，全身洁白并泛有一层金属质感的银色反光，极具高贵典雅之气。红顶虎与人的亲和度互动性非常好，因此该品种备受欢迎。为了表达对许祺源先生的尊敬，大家也把它称为许氏鹅头红。以前红顶虎的正品率较低，不过近些年来，几代金鱼业者通过不断努力，使得红顶虎的正品率大大提高。

Red Cap Lionhead (Northern) is also called Cat-Goose in Wuhan area. It is the most remarkable star in Lionhead family. The goldfish predecessors, Mr. Xu Qiyuan from Wuxi, Jiangsu, and Mr. Ma Baogen a skilled goldfish breeder in Suzhou, through years of target cultivation, hybrid cultivate Red Cap individuals in Red White Lionhead (Northern) and Lionhead (with Well Developed Head Growth and Chubby Body) and continuously purify to obtain Red Cap Lionhead (Northern). According to its head features it is more like Wuhan Lionhead (with Well Developed Head Growth and Chubby Body). The head grows in horizontal trend and sideburns pompon and gill pompon are well developed. On the middle top of the head is embedded in a square of red mark. It is middle-sized and it has a totally white body with metallic shiny silver reflection full of elegance. Red Cap Lionhead (Northern) is easy to close and good at interaction with people so it is very popular. People call it Xu style of Red Cap Goosehead to show their respect to Mr. Xu Qiyuan. In the past Red Cap Lionhead (Northern) genuine rate is low. In recent years, through continuous efforts of several generations of goldfish industry, the Red Cap Lionhead (Northern) genuine rate has greatly increased.

黑白长尾虎头

Black White Lionhead
Long Tail (Northern)

- **花色类别** / Color

  黑白 / Black White

- **品种类别** / Species

  蛋形 > 头型变异类 > 狮头型

  Egg-Fish > Head Variation > Lionhead

- **品种详述** / Species Description

  　　图例所示为黑白长尾虎头，是在蓝虎头和蓝蛋凤杂交选育的蓝长尾虎头中，褪色产生的一个新品种，在近些年才出现。黑白长尾虎头是一种小众品种，色彩稳定性不强，通过对温度及水质的控制，可以延长其观赏期。长尾虎头是由虎头和蓝蛋凤杂交选育的，因此体形一般比虎头偏长一些，头茸比较适中，不会显得过分夸张。

  As shown in the illustration is Black White Lionhead Long Tail (Northern), occurred in recent years. It is a new breed from color-fading, by hybrid cultivation of Blue Lionhead and Blue Egg-Fish Phoenix Tail. Black White Lionhead Long Tail (Northern) is a small variety. Its color stability is not strong. The viewing period can be extended by controlling the temperature and water quality. Since it is hybrid cultivated from Blue Egg-Fish Phoenix Tail, its size is a little bit longer than Lionhead and its head pompon is in middle size which will not seem too exaggerated.

麒麟虎头 — Kirin Lionhead (Northern)

- **花色类别** / Color
  软鳞红黑 / Kirin

- **品种类别** / Species
  蛋形 > 头型变异类 > 狮头型
  Egg-Fish > Head Variation > Lionhead

- **品种详述** / Species Description

　　图例所示为麒麟虎头。它是南方风格的狮头型虎头，系从五花虎头中选育出。其头茸发达但颗粒大而松，横向和纵向取势较为均衡，顶茸与两侧头茸分界线不太明显，方头型与圆头型兼有。麒麟虎头是软鳞五花虎头中出现全身青灰色或青黑色的个体，其鳞片颜色一般基部较深，边缘处较浅，排列整齐，十分富有层次感。麒麟花色如果配以红顶者，会显得更加高贵。

As shown in the illustration is Kirin Lionhead (Northern), which is Lionhead of the southern style cultivated from Calico Lionhead (Bluish Base) (Northern). Its head pompon is well-developed with big and loose particles. The pompon is well distributed vertically and horizontally. The dividing line between head pompon and both sides head pompon is not very clear. It has both square head and round head. Kirin Lionhead (Northern) is a individual with blue-gray or blue-black among Calico Lionhead (Bluish Base) (Northern). Generally, the color of the base of the scale is usually dark, the edge is shallow. Its scales arranges in a neat, full of sense of hierarchy. If Kirin Color in collocation with Red Cap, it will looks more noble.

五花虎头 —— Calico Lionhead (Bluish Base) (Northern)

- **花色类别** / Color
  软鳞五花 / Calico

- **品种类别** / Species
  蛋形 > 头型变异类 > 狮头型
  Egg-Fish > Head Variation > Lionhead

- **品种详述** / Species Description

　　图例所示为五花虎头，它是南方风格的狮头型虎头金鱼，其头茸发达，颗粒较大，横向和纵向取势较为均衡，顶茸与两侧头茸分界线不太明显，以方头型居多。狮头型五花虎头是南京、苏州和南通等地区的传统地方代表品种，以红顶白腹蓝背芝麻花的花色品种为上品，素蓝花、芝麻花的花色品种为优品，而白皮或虎皮通常会被淘汰。武汉业者引入五花虎头并与猫狮杂交选育出五花猫狮。五花虎头玩赏取素雅的颜色，红、黑、黄等色块不宜过大过多，点到即止。

As shown in the illustration is Calico Lionhead (Bluish Base) (Northern), which is a Lionhead of the southern style. Its head pompon is well-developed with big particles. The pompon is well distributed vertically and horizontally. The dividing line between head pompon and both sides head pompon is not very clear. Most of them are square head. Calico Lionhead (Bluish Base) (Northern) is traditional local representative of Nanjing, Suzhou, and Nantong area. Calico Lionhead (Bluish Base) (Northern) with red head, white belly, blue back with sesame flower is the best, plain blue flower and sesame flower is the excellent. While the white skin and the tiger skin will be eliminated. Calico Lionhead (Bluish Base) (Northern) introduced by Wuhan goldfish industry staff to be hybrid cultivated with Lionhead (with Well Developed Head Growth and Chubby Body) to producing Calico Lionhead (with Well Developed Head Growth and Chubby Body). Calico Lionhead (Bluish Base) (Northern) with plain color is more suitable for viewing. Red, black and yellow stains are supposed not too big and too much.

花色类别 / Color •

红 / Red

品种类别 / Species •

蛋形 > 头型变异类 > 狮头型

Egg-Fish > Head Variation > Lionhead

品种详述 / Species Description •

红寿星 —— Red Lionhead (Southern)

　　寿星是福州的传统地方代表品种之一，其头茸发达，颗粒较大，头茸发育以纵向取势，顶茸与两侧头茸分界线不太明显，浑然一体，以方头型居多，寿星头茸分布符合蛋形金鱼头型变异类中狮头型的定义。寿星顶茸发育较高，如同神话传说中大额头的南极仙翁，所以称之为寿星。因此，家里能饲养一缸福州产的寿星，自然就有了福寿双全的美好寓意。红寿星通体红色，以平背小尾者为上品。

Lionhead (Southern) is traditional representative breed of Fuzhou. Its head pompon is well developed with big particles, growing in vertical trend. The dividing line between head pompon and both sides head pompon is not clear and they integrated together. The head of most Lionhead (Southern) are square head. Lionhead (Southern) head pompon distribution is coincide with Lionhead definition of Egg-Fish Head Variation. Its head pompon grows high, just like the supernatural being in the South Pole in the legend who has a large forehead. That is why it named Lionhead. Therefore, breeding a cylinder of Lionhead Goldfish from Fuzhou at home has a good moral that enjoy both happiness and longevity. Red Lionhead (Southern) is totally red, the plain back with small tail is the best.

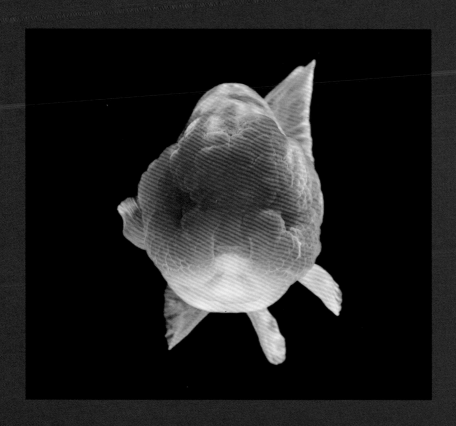

花色类别 / Color •

红白 / Red White

品种类别 / Species •

蛋形 > 头型变异类 > 狮头型

Egg-Fish > Head Variation > Lionhead

品种详述 / Species Description •

　　红白寿星由红寿星选育而出，其外部形态特征与红寿星完全一样。通过福州几代金鱼业者的努力，红白寿星的花色组合不断变化。最著名的是如图所示的首尾红寿星，其全身素白洁净，没有一丝杂色，在齐鳃红的基础上又叠加了全红尾，首尾呼应，相得益彰，可遇不可求。

Red White Lionhead (Southern) is a result of selectively breeding from Red Lionhead (Southern). Its body characteristics are the same as those of the red. Decades of hard work by the Fuzhou goldfish industry has lead to a continuous change in the color pattern of the Red White Lionhead (Southern). The most famous is as shown in the illustration Head and Tail Red Lionhead (Southern) . Its whole body is spotless white, without a trace of other colors. On the basis of red gill the tail is red. the head and tail bring out the best in each other, one can not ask for more.

红白寿星
Red White Lionhead
(Southern)

红顶寿星

Red Cap Lionhead (Southern)

- **花色类别** / Color
  红白 / Red White

- **品种类别** / Species
  蛋形 > 头型变异类 > 狮头型
  Egg-Fish > Head Variation > Lionhead

- **品种详述** / Species Description

　　红顶寿星和红顶虎头如同孪生兄弟一般，不容易分清楚。二者有相同的审美意趣，区别在于红顶寿星头茸纵向取势，红色顶茸较高；而红顶虎头因有猫狮的血缘，头茸横向取势，红色顶茸不凸出，甚至有点如同凹陷般的镶嵌感。由于地区差异及饲养手法不同，红顶寿星体高，身形较壮实，俯视侧视皆可，而红顶虎头偏向于俯视。红顶寿星由红白寿星不断筛选定型，遗传稳定性较好，是福州金鱼不可多得的优秀品种之一。

The Red Cap Lionhead (Southern) and the Red Cap Lionhead (Northern) are almost like twins, very hard to tell apart. They have the same wonderful qualities. The difference lies in that the Red Cap Lionhead (Southern)'s head pompon grows vertically and the red head pompon is high. Because Red Cap Lionhead (Southern) has a blood relationship with Lionhead (with Well Developed Head Growth and Chubby Body) its head pompon grows horizontally, and the head pompon does not protrude, giving it a sunken in look. Because of the differences in local differenced and breeding practices, Red Cap Lionhead (Southern) has a tall robust body and overlook and side looking are both ok. While Red Cap Lionhead (Southern) is more suitable to over look. Continuous selective breeding of the Red White Lionhead (Southern) has given the Red Cap Lionhead (Southern) a good hereditary stability. It is one of the excellent breeds to come out of the Fuzhou goldfish industry.

虎纹寿星 | Tiger Banded Lionhead (Southern)

- **花色类别** / Color

  软鳞红黑 / Tiger Banded

- **品种类别** / Species

  蛋形 > 头型变异类 > 狮头型

  Egg-Fish > Head Variation > Lionhead

- **品种详述** / Species Description

  虎纹寿星是从五花寿星中选育出的软鳞花色品种。过去，虎纹花色一直作为五花花色中的次品而被淘汰，但近些年来，虎纹花色在东南亚一带逐渐流行并影响到国内市场。虎纹寿星外部形体特征和其他寿星金鱼完全一样，头茸非常发达。生长发育期应适当控制，防止二龄后头茸开花松散。虎纹花色应取橙黄色或橙红色底色，从背部延伸至腹部的黑色条状斑纹，应疏密有度，如同老虎的花纹。

  Tiger Banded Lionhead (Southern) is a matt breed and it is the result of selectively breeding from Calico Lionhead (Bluish Base). In the past, Tiger banded color was always seen as substandard and eliminated through selection. But in recent years it has become more and more popular in southeast Asia which has influenced China's domestic market Tiger Banded Lionhead (Southern)'s body characteristics are the same as other lionhead goldfish's head pompon, extremely well-developed. Proper control during the growth phase can prevent the head pompon from scattering after it has reached two years of age. The background color of tiger striped color should be orange yellow or orange, extending from the back to the abdomen, forming stripes with the same spacing and angles as the pattern on a tiger.

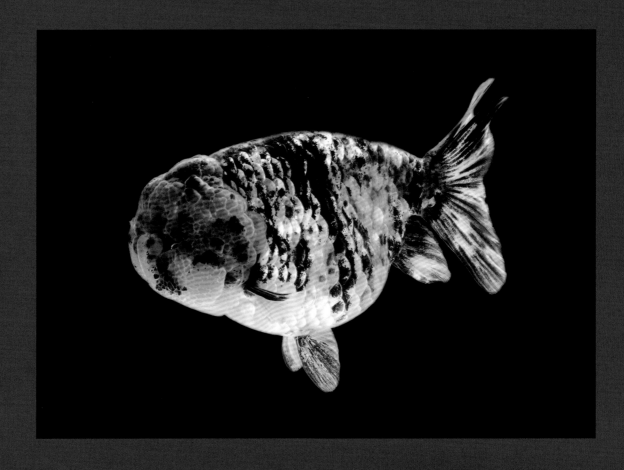

花寿星 | Calico Lionhead (Southern)

- **花色类别** / Color
  软鳞五花 / Calico

- **品种类别** / Species
  蛋形 > 头型变异类 > 狮头型
  Egg-Fish > Head Variation > Lionhead

- **品种详述** / Species Description

　　同虎纹寿星一样，花寿星也是从五花寿星中选育出的软鳞花色品种，其花色介于五花和虎纹之间，从传统金鱼审美角度而言，这种花色和虎纹一样，在遗传中居于强势稳定的地位，属于五花中非主流的审美类型，如果想选育五花后代，应避免选择这样的花色品种作为亲本。图例所示的花寿星体态雄健，头茸发达。其体质较强健，易于饲养，与人的亲和度和互动性都较好，属于较受欢迎的普通商品金鱼品种。

Like the Tiger Banded Lionhead (Southern), the Calico Lionhead (Southern) is also a matt breed that is the result of selectively breeding Calico Lionhead (Bluish Base). Its color pattern is a mix between the Calico (Bluish Base) and the Tiger Banded. From the point of view of traditional standards of beauty, the same as Tiger banded, this type of color pattern is in strong and stable position genetically, which is non-mainstream standards of beauty. If one wants to breed Calico(Bluish Base) descendants it is best to avoid using breeding stock which has this type of color pattern. Calico Lionhead (Southern) shown in the illustration, of which the body is vigorous and the head pompon is well-developed. Its physique is fairly strong and healthy, and it is easy to rear. It is good at interactions with people and very easy to close to. It is a fairly popular and common goldfish.

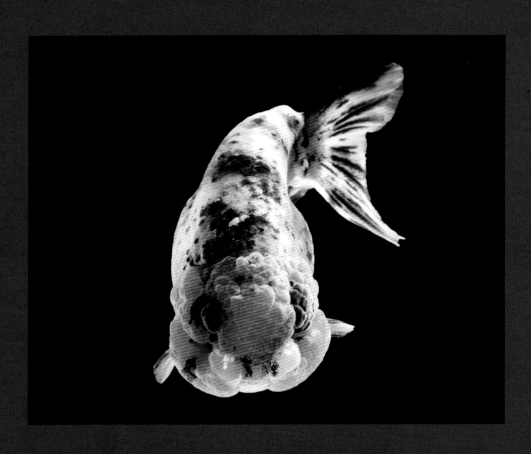

红兰寿
Red
Ranchu

- **花色类别** / Color

  红 / Red

- **品种类别** / Species

  蛋形 > 头型变异类 > 狮头型

  Egg-Fish > Head Variation > Lionhead

- **品种详述** / Species Description

　　中国兰寿又称国寿或福寿，是东莞的金鱼业者将日寿进入中国后，由福州的金鱼业者按照中国人的审美习惯，逐步培育出的一个品种。目前的日寿和国寿风格迥异，但回溯到三四十年前，当时的日寿与现在的国寿，风格上并无很大的差异。大多数国寿，在头茸分布上符合金鱼狮头型的定义。20世纪30年代，许和所著的《金鱼丛谈》中有一幅狮头的照片，其小尾高体弓背的侧视外形，和今天的国寿十分近似。红兰寿通体红色，体质强健，易于饲养，但应注意背弧线最好为长抛物线，后半背弧不宜收得过急，否则易倾头。

Chinese Ranchu, also known as National Ranchu or Fuzhou Ranchu, is a breed cultivated by the goldfish industry in Fuzhou through progressively breeding Japanese Ranchu according to the tastes of the Chinese after Japanese Ranchu had been introduced into China by the goldfish industry in Dongguang. At the present, the styles of the Japanese Ranchu and National Ranchu are vastly different. But back to thirty or forty years ago, there were no so much differences between the Japanese Ranchu and national Ranchu at present in style. In the definition of head velvet distribution, most National Ranchu is in accordance with Tigerhead. In the 20th century and 30 years, there was a picture of Lionhead in the book *On Goldfish* by Mr. Xu He, viewed from the side, its small tail, tall body and arched back give it a very similar appearance to the red National Ranchu we have today. The Red Ranchu has a totally red body and a strong and healthy physique and is easy to rear. But to be noted that the back arc curve is better to be parabola and last half part of back should not be too hasty, otherwise the head is prone to dumping.

黑兰寿 —— Black Ranchu

- **花色类别** / Color
  黑 / Black

- **品种类别** / Species
  蛋形 > 头型变异类 > 狮头型
  Egg-Fish > Head Variation > Lionhead

- **品种详述** / Species Description

　　黑兰寿在外部特征上与红兰寿完全一样，头茸多颗粒状，较发达。黑兰寿能做到通体乌黑如墨的极为稀少，通常腹部会呈青白色或青黄色。而东南亚一带饲养的黑兰寿却能做到通体漆黑如墨锭，这其中除了温度和光照等因素外，亲本金鱼的选择也不能忽视。黑兰寿在操作时亦要非常小心，黑色鳞片刮擦后会留下痕迹，需要很长时间方能恢复，大大影响其欣赏价值。黑兰寿体质较强，易于饲养。

The external characteristics of Black Ranchu are exactly the same as Red Ranchu. Its head pompon a little bit developed with many particles. Black Ranchu can be totally black as black ink. While, its belly is usually bluish white or green yellow. Black Ranchu reared in Southeast Asia area can be as black as ink ingot. That is because of factors as following: the temperature and light. Besides, choosing the parents goldfish is also very important. It has to be very careful when do the operation on Black Ranchu. Once scratched, its black scales will leave traces, and it takes a long time to recover. As a result, its appreciation value will be greatly reduced.

白兰寿
White Ranchu

- **花色类别** / Color
  白 / White

- **品种类别** / Species
  蛋形 > 头型变异类 > 狮头型
  Egg-Fish > Head Variation > Lionhead

- **品种详述** / Species Description

　　白兰寿由红白兰寿中选育而出，其外形特征和红兰寿没有太大区别。纯粹的白兰寿欣赏价值不高。但白兰寿中常有带红色眼睛的个体，称为朱砂眼或玉兔，颇具欣赏价值。白兰寿由于受到光照、水质及其他因素的影响，头部和尾部有时会显现深浅不同的柠檬黄色，图例所示的白兰寿便是如此。其鳞片光洁银亮，纯洁无瑕，背弧曲线优美，配以淡柠檬黄的头茸和尾鳍更显典雅高贵，而朱砂色的兔眼略略增加了一丝俏皮。

White Ranchu is cultivated from Red White Ranchu. There is no so much difference in external characteristics between White Ranchu and Red White Ranchu. The appreciation value of Red White Ranchu is not high. But there is some individuals of White Ranchu with red eyes, known as cinnabar eyes or rabbit has high appreciative value. Impacted by some factors such as the light, water quality and some other factors, Its head and tail sometimes are lemon yellow different in shade, just as the White Ranchu shown in the illustration. Its smooth silver scales bright and shiny, pure and flawless; beautiful back arc curve with pale lemon yellow head pompon and caudal fins are more elegant an noble, and cinnabar-colored rabbit eyes slightly increase the hint of playful.

蓝兰寿
Blue Ranchu

- **花色类别** / Color
  蓝 / Blue

- **品种类别** / Species
  蛋形 > 头型变异类 > 狮头型
  Egg-Fish > Head Variation > Lionhead

- **品种详述** / Species Description

　　图例所示为蓝兰寿，背部弧线曲度优美，游姿稳健，静止时不倾头。其外形特征和红兰寿没有太大区别。蓝兰寿大多为墨蓝色，鳞片基部青黑色，边缘部分带有银色反光，折射出淡蓝色的光泽。蓝兰寿是近十年选育出的国寿新品种，其中又会褪变出黑白色或喜鹊花色，亦会出现黛斑鳞，但都不太稳定。蓝兰寿的头茸发育度不及其他的花色类型，以侧视观赏为主。在水族箱中配以合适的灯光，才能尽显其美。

As shown in the illustration is Blue Ranchu, with beautiful back arc curve. Its swimming posture is steady, and the head is not inclined when it is resting. There is no so much difference in external characteristics between Blue Ranchu and Red White Ranchu. Most Blue Ranchu is dark blue, scale base black, the edge of the part with a silver reflection, reflecting the light blue luster. Blue Ranchu has been a new breed of Chinese Ranchu in recent 10 years, when it will fade out black and white or magpie flower color and sometimes it will have lividity fleck scales, all of which are not stable. Its head pompon develops less than that of other colors of the variety. It mainly view from the side. Put proper light in the aquarium in order to show its beauty fully.

紫兰寿

Chocolate Ranchu

- **花色类别** / Color
  紫 / Chocolate

- **品种类别** / Species
  蛋形 > 头型变异类 > 狮头型
  Egg-Fish > Head Variation > Lionhead

- **品种详述** / Species Description

紫兰寿的出现略早于蓝兰寿。图例所示紫兰寿背弧曲线优美，头茸发育适中且较规整，不散乱。全身紫铜色并附有光泽，腹部鳞片排列略微有些不整齐，为饲养时未能控制好身体的生长速度所致，这使其欣赏价值有所折扣。紫色鳞片和黑色鳞片一样，操作时亦要非常小心，其鳞片刮擦后会留下痕迹，需要很长时间方能恢复。紫兰寿亦有紫绶金章的含义，比较受大家的欢迎。

The occurrence of Chocolate Ranchu is a little bit earlier than that of Blue Ranchu. As shown in the illustration is Chocolate Ranchu which has beautiful back arc curve, neat and middle-delveloped head pompon, not scattered. Its body is copper-colored and shiny. The belly scales arrangement is slightly irregular, because the growth rate is not well-controlled. As a result , the appreciative value will be reduced. Like the black scales the Chocolate scales also should be very carefully in operation. Once scratched, its scales will leave traces, and it takes a long time to recover. The Chocolate Ranchu also has the meaning of "Zi Shou Jin Zhang" (be honored with high official titles) so it is popular with people.

雪青兰寿
—— Lilac Ranchu

- **花色类别** / Color
  雪青 / Lilac

- **品种类别** / Species
  蛋形 > 头型变异类 > 狮头型
  Egg-Fish > Head Variation > Lionhead

- **品种详述** / Species Description

　　雪青兰寿的出现基本和蓝兰寿出现的时期相近或略晚，是从紫兰寿中选育而出。图例所示雪青兰寿背弧曲线优美，头茸发达且规整，顶茸高耸，与两侧头茸分界明显，和雪青虎头的三块瓦头型颇为类似。雪青兰寿全身无杂色，鳞片排列规整，富有金属质感光泽。雪青兰寿对水质及光照的要求相对其他花色的国寿要略高一些，否则其色彩表现会有所折扣。雪青兰寿遗传较为稳定，但为小众品种，存世量不多，亦属于珍稀金鱼品种。

The occurrence of Lilac Ranchu is close to or a little bit later than that of Blue Ranchu, and it is cultivated from Chocolate Ranchu. As shown in the illustration is the Lilac Ranchu which has beautiful arc curved back and developed head pompon, very neat. There is clear dividing line between the high head pompon and the side velvet which is similar as the three watt head type of Lilac Lionhead. Its scales arrangement is neat and regular full of metallic luster. Its body is pure and non-variegated. Compared with other color variety in Chinese Ranchu. Lilac Ranchu is stricter to water quality and light, or its appreciative value will be reduced. It is genetic stable and a mall variety which is not so much in the world so it is also a rare breed of gold fish. Lilac Ranchu genetic stable, but for the minority species, quantity is not much. It also belongs to the rare varieties of goldfish.

红白兰寿 │ Red White Ranchu

- **花色类别** / Color
  红白 / Red White

- **品种类别** / Species
  蛋形 > 头型变异类 > 狮头型
  Egg-Fish > Head Variation > Lionhead

- **品种详述** / Species Description

　　红白兰寿是福寿各种花色品种中，最受欢迎的一个花色品种。日寿中一般赤胜更纱的品味要高于白胜更纱，而国内的审美习惯是白胜更纱更受欢迎，这是两国文化的差异体现。图例1中的红白兰寿是白胜更纱的一种首尾红。从外部形态看，在尾柄处略微收得局促。图例2中是赤胜更纱，其头茸紧实有型，不松散。身形丰满，雍容大度，是国寿的典型代表品种。

Among various color varieties of Fuzhou Ranchu, Red White Ranchu is the most popular variety. As for Japanese Ranchu, the taste of Akagachisarasa is higher than that of the Shirogachisarasa. While in china, the Shirogachisarasa is more popular which reflects the cultural differences between China and Japan. As shown in the illustration 1 is a type of Shirogachisarasa with red head and tail. From the external form, the caudal body ends up too fast. As shown in the illustration 2 is Akagachisarasa. Its head pompon is tight and in good shape, non-scattered. Its physique is plump and generous and it is a typical representative breed of Chinese Ranchu.

花色类别 / Color •

紫红 / Chocolate Red

品种类别 / Species •

蛋形 > 头型变异类 > 狮头型
Egg-Fish > Head Variation > Lionhead

品种详述 / Species Description •

紫红兰寿 —— Chocolate Red Ranchu

　　图例所示为紫红兰寿，其外形特征和红兰寿没有太大区别，是紫兰寿向红兰寿转色的过渡期品种，颜色表现基本不稳定。许多紫色金鱼都有这样的现象。义形金鱼中的著名品种朱顶紫罗袍也是这种变化品种。通过人为控制水质和温度的变化，或在饲料中添加一些螺旋藻，可以延缓这样的变色过程。金鱼业者是否可以通过定向选育的方式将紫红色稳住还不得而知，但目前在文球和龙睛球中已经实现。

As shown in the illustration is Chocolate Red Ranchu. There is not so much difference in external characteristics between Chocolate Red Ranchu and Red Ranchu. It is a variety during the color transition period of that Chocolate Ranchu turns into Red Ranchu. Basically, the color is unstable. Most Chocolate Goldfish is like that. The famous variety of Red Head with Chocolate Luopao Body in Fantail Goldfish is also this type of variation. The color transition process can be extended by artificially controlling the water quality and temperature changes or adding some spirulina. Whether the Chocolate and red color can be stabilized by directive cultivation or not, we don't know. But it can be realized in Fantail Goldfish with Pompons and Telescope with Pompons by goldfish industry.

黑白兰寿
Black White Ranchu

- **花色类别** / Color
  黑白 / Black White

- **品种类别** / Species
  蛋形 > 头型变异类 > 狮头型
  Egg-Fish > Head Variation > Lionhead

- **品种详述** / Species Description

　　图例所示为黑白兰寿。黑白兰寿是从蓝兰寿中选育而出，通常是蓝兰寿褪色时期的一个过渡期品种，一般很难达到比较稳定的黑白色。但是，我们可以通过人为地控制水质、温度以及在饵料中添加螺旋藻等方法延缓褪色的过程。黑白兰寿外部形态和其他中国兰寿无太大区别，体形中等。黑白兰寿以侧视欣赏为主，可以和其他花色的品种混合搭配，并辅助以合适的灯光，从而达到最佳欣赏效果。

As shown in the illustration is Black White Ranchu which is cultivated from Blue Ranchu. Usually it is a variety during the color fading transition period of Blue Ranchu and it is normally very hard to obtain stable colors of Black White. But the color transition process can be extended by artificially controlling the water quality and temperature changes or adding some spirulina. Its external characteristics are not so much different from that of other Chinese Ranchu. It is in middle size. Black White Ranchu is mainly suitable for side viewing, collocated with other colors varieties, assisted with proper lighting, so as to achieve the best appreciation effect.

花色类别 / Color •

紫蓝 / Chocolate Blue

品种类别 / Species •

蛋形 > 头型变异类 > 狮头型

Egg-Fish > Head Variation > Lionhead

品种详述 / Species Description •

　　图例所示为紫蓝兰寿。紫蓝兰寿一般是从紫兰寿和雪青兰寿中选育而出，在蓝兰寿中偶尔也会出现。紫蓝兰寿是比较小众的一个花色品种，饲养的也不是很多。图例所示的两尾紫蓝兰寿，化色规整，鳞片整齐，光泽度好。外部体态上，其头茸发育良好，憨态可掬，背部圆顺，弧线优美，呈长抛物线状，是形色俱佳的国寿精品。

As shown in the illustration is Chocolate Blue Ranchu which is usually cultivated from Chocolate Ranchu and Lilac Ranchu, occasionally from Blue Ranchu. It is a relatively small minority of the varieties. There is not so many breeders. As shown in the illustration are two Chocolate Blue Ranchu with regular colors, neat scales and nice luster. From the external characteristics, it has well-developed head pompon, charming naivety, smooth round back, beautiful arc curve, long parabolic shape. It is a competitive goldfish of Chinese Ranchu with good color and taste.

紫蓝兰寿 — Chocolate Blue Ranchu

三色兰寿 ｜ Red Black White Ranchu

- **花色类别** ／ Color
  三色 ／ Red Black White

- **品种类别** ／ Species
  蛋形 ＞ 头型变异类 ＞ 狮头型
  Egg-Fish ＞ Head Variation ＞ Lionhead

- **品种详述** ／ Species Description

　　三色兰寿多为硬鳞花色，常见的是黑白红三色和紫白红三色。和黑白兰寿类似，三色兰寿也多为过渡色品种，我们同样也可以通过人为地控制水质、温度以及在饵料中添加螺旋藻等方法延缓其褪色的过程。三色兰寿的体色一直处于变化的过程，一般在冬季和春季达到颜色最佳的状态，是欣赏的最佳时期。一些红白兰寿在繁殖期，由于体内激素的作用，会在体表出现泛色现象，变为红白黑三色，但繁殖期结束后又会恢复。三色兰寿最终会变为红白兰寿，但其子代中还会有三色个体出现。

Most Red Black White Ranchu are hard scales calico. The common three-colors are two types : black, white, red and chocolate, white, red. Similar as Black White Ranchu, Red Black White Ranchu is also a color transition variety. The color transition process can also be extended by artificially controlling the water quality and temperature changes or adding some spirulina. The color of its body is always in a process of change. Usually in winter and spring the color reaches the best state and it is the best time to view. During the breeding period, because of the hormone, Anti-color phenomenon occurs in the surface of some Red Black White Ranchu and the colors turn into red, white, and black. It will recover after its breeding season. Red Black White Ranchu will eventually turn into Red White Ranchu, but there will be also Red Black White individuals among its offspring.

樱花兰寿

Sakura Ranch
(Red White Matt)

- **花色类别** / Color
  软鳞红白 / Red White Matt

- **品种类别** / Species
  蛋形 > 头型变异类 > 狮头型
  Egg-Fish > Head Variation > Lionhead

- **品种详述** / Species Description

　　图例所示为樱花兰寿，其外形特征和五花兰寿没有太大区别。樱花色是指软鳞的红白花色，并且在金鱼体表上，点缀少许白色的反光鳞片，如飘落的片片樱花花瓣。在日本金鱼中也有类似的花色品种称之为樱锦，而不称之为兰寿。樱花兰寿体表的反光鳞不宜过多，所谓点到即止。樱花兰寿的头茸发育度要弱于硬鳞的红白兰寿，但身形可以放得很大，这正体现了中国人所说的"失之东隅，收之桑榆"的哲学思想。

As shown in the illustration is Sakura Ranch (Red White Matt). Its external characteristics are not so much different from that of Calico Ranchu (Bluish Base). Red White Matt is soft scale Red White, dotted with a few white reflective scales on the surface, as falling petals of cherry blossoms. In Japan, there is also similar color pattern variety named Sakura Nishiki, but not Ranchu. The reflective scales on the Sakura Ranchu (Red White Matt) surface should not be too much, a few is enough. Sakura Ranchu (Red White Matt)' s head pompon development is not as good as that of Red White Ranchu with hard scales. But it can grow into big size, which is in accordance with a Chinese philosophy "When God closes a door in front of you, he will open one more window beside you".

云石兰寿

Marble Ranchu

- **花色类别** / Color
  软鳞黑白 / Marble

- **品种类别** / Species
  蛋形 > 头型变异类 > 狮头型
  Egg-Fish > Head Variation > Lionhead

- **品种详述** / Species Description

　　图例所示为云石兰寿。云石兰寿是近几年刚刚诞生的一个新的花色品种，是金鱼业者从五花兰寿中选育的。云石兰寿为软鳞鱼，身上有一些硬质反光鳞，在灯光下欣赏有非常炫目的效果。软鳞黑白色不同于硬鳞黑白色，是比较稳定的颜色，通常不会褪色。图例所示金鱼色彩、体形以及头茸都恰到好处，是非常优秀的一尾国寿代表品种。

As shown in the illustration is Marble Ranchu, which is a new variety, just born in recent years. It is cultivated from Calico Ranchu (Bluish Base) by goldfish industry. Marble Ranchu is a kind of soft scale fish, with hard reflective scales on its body which is very dazzling under the light. Unlike the black and white color of hard scales, the black and white colors of soft scales is relatively stable, normally non-fading. The color, size and head pompon of the goldfish shown in the illustration are just perfect and it is a very excellent representative variety of Chinese Ranchu.

水墨兰寿 | Water Pattern Ranchu

- **花色类别** / Color
  软鳞黑白 / Water Pattern

- **品种类别** / Species
  蛋形 > 头型变异类 > 狮头型
  Egg-Fish > Head Variation > Lionhead

- **品种详述** / Species Description

　　图例所示为水墨兰寿。水墨兰寿也是近几年刚刚诞生的一个新的花色品种，是金鱼业者从五花兰寿中选育的。水墨兰寿由于是软鳞，其花色表现效果如同生宣纸上水墨晕染的效果，又像大理石的肌理效果，让人感叹造化的神奇。水墨兰寿鳞片上消耗的钙磷等较硬鳞鱼少，这可以让骨骼充分发育，所以一般软鳞花色的鱼都较硬鳞花色的鱼体形大。

As shown in the illustration is Water Pattern Ranchu, which is also a new variety, just born in recent years. It is cultivated from Calico Ranchu (Bluish Base) by goldfish industry in Fuzhou. Because of its soft scales, the color performance is as that of the ink on rice paper, and also like a marble texture, Which makes people sign with good feeling about the Magic of nature. The consumption of calcium and phosphorus on Water Pattern Ranchu scales is less than that on hard fish scales, which can promote the full development of bone. So normally the size of soft scales fish are bigger than that of hard scales.

虎纹兰寿
Tiger Banded Ranchu

- **花色类别** ／ Color
  软鳞红黑 ／ Tiger Banded

- **品种类别** ／ Species
  蛋形 ＞ 头型变异类 ＞ 狮头型
  Egg-Fish ＞ Head Variation ＞ Lionhead

- **品种详述** ／ Species Description

　　虎纹兰寿是从五花兰寿中选育的软鳞花色品种。虎纹兰寿外部形态特征和其他五花国寿完全一样，头茸适中，体形雄健，饲养得法，可以成长为30厘米以上的巨寿。虎纹花色应取橙黄色或橙红色作底色，黑色条状斑纹不要块状分布，色彩的搭配从背部延伸至腹部应疏密有度，如同老虎的花纹。虎纹兰寿在观赏中只作为点缀，不可过多，一群中有一两尾即可。

Tiger Banded Ranchu is cultivated from Calico Ranchu (Bluish Base). Its external characteristics are totally the same as that of other Calico Chinese Ranchu. It is vigorous with moderate head pompon. If well raised it can grow up to a Giant Ranchu over 30 cm. As Tiger Banded, the background color should be orange-yellow or orange-red, black tripe lines should not distributed in blocks and the color collocation from the back extending to the belly should be in proper density like the tiger pattern. Tiger Banded Ranchu is only as an embellishment which should not be too much and one or two in a group is enough.

花兰寿 | Calico Ranchu

- **花色类别** / Color
  软鳞五花 / Calico

- **品种类别** / Species
  蛋形 > 头型变异类 > 狮头型
  Egg-Fish > Head Variation > Lionhead

- **品种详述** / Species Description

　　花兰寿一般为五花兰寿中分离出的一个花色品种，通常将没有蓝色基色的软鳞多花色鱼归类在其中。由于是软鳞花色鱼，所以可以养得比较硕大。软鳞花色鱼的头茸发育比硬鳞花色的要稍逊一筹，但一般比较规整，不易开花松散。花兰寿身上的墨和绯，随着成长时间会加深，而淡蓝色或淡青色的翠色会逐渐丢失。花兰寿体质强健，较容易饲养，是比较受欢迎的商品级兰寿。

Generally, Calico Ranchu is a variety isolated from Calico Ranchu (Bluish Base). Those various Calico Ranchu without primary colors of blue is pigeonholed as Calico Ranchu. Because it is calico, it can grow in big size. The head pompon development of soft scale fish is not so good as that of hard scale Goldfish, but relatively neat, not easy to scatter. The black and red color of Calico Ranchu will deepen with the growth time, while, the pale blue or pale blue cruise will gradually fade away. With strong physique, Calico Ranchu is easy to rear, and it is a popular commercial grade Ranchu.

重彩兰寿 —— Ranchu (with Intense Colors)

- **花色类别** / Color
  软鳞五花 / Intense Colors

- **品种类别** / Species
  蛋形 > 头型变异类 > 狮头型
  Egg-Fish > Head Variation > Lionhead

- **品种详述** / Species Description

　　重彩兰寿也是金鱼业者近些年来，从五花兰寿中分离出来的一个新的花色品种，其身上的墨和绯都比较浓重。和虎纹兰寿类似，这种花色品种在东南亚比较受欢迎。重彩兰寿和水墨兰寿展现着不同的中国风，其给人的总体感觉是强烈的激情和热烈的奔放，又有一种印象派油画的光影效果。重彩兰寿一般都是体形硕大的巨寿，给人强烈的视觉冲击力，如同征战沙场的武士，"铁盔金甲朱龙绦，百战敌血染征袍"。因此，近几年，在国内外的各种金鱼大赛中，重彩兰寿获奖颇丰，是倍受欢迎的高档国寿品种。重彩兰寿特别适合在玻璃水族箱中观赏，配以适合的顶光，可以将其如同张大千先生笔下泼彩水墨山水画的中国风完美地诠释出来。

Ranchu (with Intense Colors) is also a new variety in recent years, cultivated from Calico Ranchu (Bluish Base) by goldfish industry in Fuzhou. The black and red on its body are strong. Similar with Tiger Banded Ranchu, it is popular in East-southern Asia. The Chinese style of Ranchu (with Intense Colors) is different from that of Water Pattern Ranchu. Ranchu (with Intense Colors) explains the feeling of warm intense and unrestrained passion, and also has the light and shadow effects impressionist painting. Ranchu (with Intense Colors) is a giant life in huge size, giving people a strong visual impact and it just like the warrior in the battlefield, therefore, it has wined many prices in various goldfish contests at home and abroad. It is the most popular and high-end variety of Chinese Ranchu. It is especially suitable for viewing in the glass Aquarium with proper light on the top and it is the perfect interpretation of Chinese style as the splash of color ink landscape painting by Mr. Zhang Daqian.

五花兰寿
Calico Ranchu (Bluish Base)

- **花色类别** / Color

  软鳞五花 / Calico (Bluish Base)

- **品种类别** / Species

  蛋形 > 头型变异类 > 狮头型

  Egg-Fish > Head Variation > Lionhead

- **品种详述** / Species Description

　　五花兰寿是国寿大家族中一个传统的花色品种，也一直广受欢迎。图例1是比较标准的五花花色——红顶蓝背白腹芝麻花，其周身比较素净，风格与浓烈奔放的重彩兰寿不同，属于典雅型。从选育的角度而言，五花兰寿能达到标准花色的比例少之又少，不仅和亲本种鱼有关，同时也和产卵孵化期的水质、温度、温差等诸多因素都有关系。出苗后的饲养密度、生长速度、转色期的光照强度以及饲料中各微量元素的配比，对其花色的形成都有影响。可见，育成一尾优秀的五花兰寿是十分不易的。五花兰寿背部线条一定要舒展放开，这样才有一定的生长空间，这是由其骨骼形态所决定的，后天很难改变。

Calico Ranchu (Bluish Base) is a traditional variety in Chinese Ranchu (Dragonhead Type) family, and it is always widely popular. In the illustration 1 is relatively standard Calico color pattern: red head, blue back, white belly with sesame flower and sober body. Its style is elegant, different from the strong bold and unrestrained style of Ranchu (with Intense Colors). From the selective breeding point of view, the proportion of achieving standard Calico color pattern is very few, which has something to do with the parent species, as well as many other factors such as the water quality, temperature and temperature differences during the spawning and incubation period. The color's formation is affected by the following factors: the feeding density after occurrence, the growth pace, the light intensity during color transition, and the feeding ratio of each trace element in feedstuff. Therefore, it is not so easy to cultivate a good Calico Ranchu (Bluish Base). Its back line must stretch open, so that there is enough room for growth. It is decided by the skeletal morphology, and is very difficult to be postnatal changed.

蛋形
Egg-Fish

∨

头型变异类
Head Variation

∨

龙头型
Dragonhead

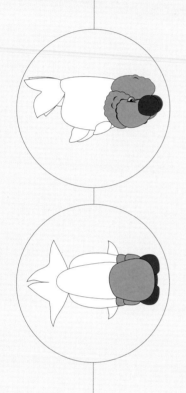

　　蛋形金鱼头型变异类以前只分3种，而当以日寿为代表的国外金鱼大量引入国内繁殖饲养后，以往的3个分类已经无法完全覆盖这些新品种，因此我们重新定义出了龙头型这个类别。日寿对头型的分类为四大类，即龙头型、狮头型、兜巾型和女假面型，而我们只借用了龙头型这个名称，但对形态的描述和它还是有所区分的。在蛋形金鱼中，我们把类似日寿和泰寿这种龙吻特别发达，头型四方的头型变异类称之为龙头型，龙头型顶茸为方形，发育度适中，鬓茸发达，尤其龙吻外凸，十分显眼，鳃茸较薄，这一点与虎头型有些接近，俯视整体呈四方形。金鱼爱好者对龙头型头茸的发育要求较高，要求规整紧实，如雕刻出来的感觉，绝对不能松散、杂乱无章。

In the past, Egg-Fish Head Variation class was only 3 types. After a large scale of Japanese Ranchu, as the representatives of foreign domestic goldfish, were introduced in China to be reared and breed, conventional three categories was unable to fully cover these new varieties. So people redefined the Dragonhead category. According to the Head Variation, Japanese Ranchu can be classified into four categories: Tsugashira, Shishigashira, Tokingashira and Binbarigashira. We just borrowed the name "Dragonhead" of this type and the form description is different. In the Egg-Fish family, those variations that similar as Japanese Ranchu and Thailand Ranchu, with super-developed snout and square head are called Dragonhead. Its head pompon is square, moderately developed. Sideburns pompon is developed, and especially the snout is very conspicuous. Gill pompon is relatively thinner, which is somewhat similar with Tigerhead. When overlooking, we can see the whole body is square. Dragonhead has high requirements on head pompon development, such as, regular compaction, just like carved, non-loose or non-scattering.

日寿
Japanese Ranchu
(Dragonhead Type)

花色类别 / Color •

红、黑、红白 / Red, Black, Red White

品种类别 / Species •

蛋形 > 头型变异类 > 龙头型

Egg-Fish > Head Variation > Dragonhead

品种详述 / Species Description •

　　日寿是日本金鱼的代表之作，在日本有金鱼之王的美号。日寿在日本已经有百年以上的玩赏历史，目前其风格上主要分为协会系和宇野系两大流派。协会系侧重于游姿和身形的塑造，在游姿优先的基础上，不断追求头型、身形和尾型的和谐进步。而宇野系侧重于头茸及花色的观赏。图例1、3、5所示为宇野系日寿。图例2、4、6为协会系日寿。日寿的鉴赏要求其头茸发育规整，两龙吻发育适度，身形饱满雄健，其横切面应近似正方形，尾鳍张力好并富有弹性。日寿的品评在对鱼与人的亲和度上有一定的要求，在比赛中若不能保证良好状态的鱼一般不会入评。日寿除了红和红白以外，一般其他颜色是不入格的，而黑色的日寿作为红色的一个变种，是在中国繁育时得以保留下来的。因为在传统文化中，黑色即玄武，表示北方，五行主水，而水生财，因此，国人对墨色或黑色的金鱼有所偏爱。

Japanese Ranchu (Dragonhead Type) is the representative of Japanese Goldfish, known as the king of the goldfish in Japan. As pet goldfish, the Japanese Ranchu (Dragonhead Type) has more than one hundred years of history in Japan. Currently, its style is divided into two mainly schools: Kyokaikei and Unokei. The Kyokaikei focuses on the swimming posture and the body shape. On the basis of swimming posture priority, it continuously promoting the harmonious progress of the head, body shape, and the tail shape. While, the Unokei pays more attention to the appreciation of head velvet and the color. In illustration 1, 3, 5 are Unokei of Japanese Ranchu (Dragonhead Type). In illustration 2, 4, 6 are Kyokaikei. The appreciation of Japanese Ranchu (Dragonhead Type) requires that the head pompon is regularly developed; two snouts are moderately developed; the body is full of vigorous, the cross section should be approximately square, and the caudal fins should be in good tension full of elastic. The evaluation of Japanese Ranchu (Dragonhead Type) has some requirements on the affinity between fish and people. If it can not guarantee a good condition in the contest, normally, it will not be assessed. The black Japanese Ranchu (Dragonhead Type) as a red color variation, has been retained during Chinese breeding. In Chinese traditional culture, black is "Xuan Wu", which means the north. The north in Five Elements is mainly water (The Five Elements, also known as the Five Phases, the Five Agents, the Five Movements, and the Five Steps/Stages, are chiefly an ancient mnemonic device, in many traditional Chinese fields). And water can make fortune, therefore, people has preference for Black-ink color or Black Golddfish.

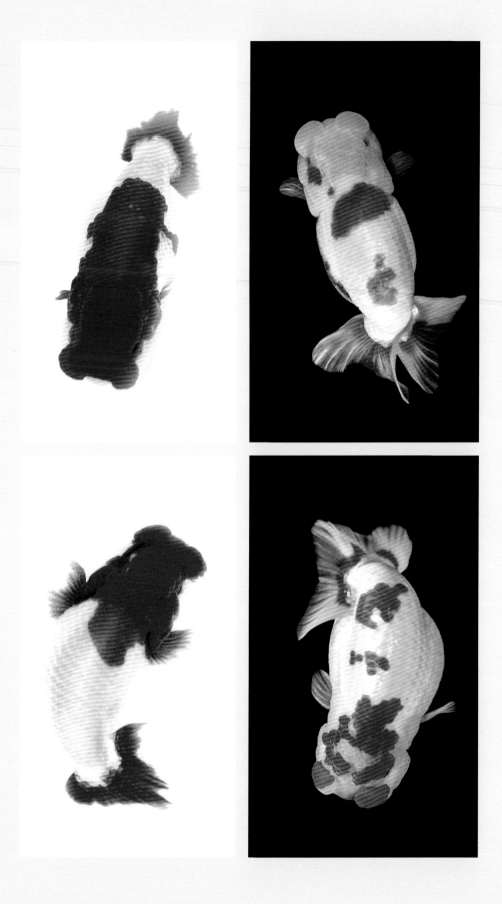

蛋形
Egg-Fish
∨
眼型变异类
Eye Variation
∨
龙睛型
Dragon Eyes

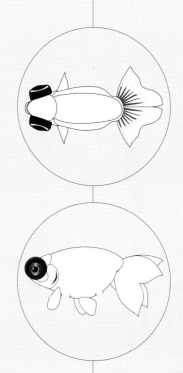

蛋形金鱼里出现眼型变异类龙睛型的品种称之为蛋形龙睛，简称为蛋龙，以往数量不多，遗传学专家曾分析过其较难培育出的遗传学原因，而现在，随着金鱼业者的不断努力，蛋形金鱼里出现龙睛型或带有龙睛型的复合变异类品种越来越多。以前，有人将蛋形金鱼里的龙睛类单独划归一类，称之为龙背，并与草、文、龙、蛋并列为第五大种，而本书不采用此种分类方法。蛋龙属于小众金鱼品种，目前还不多见，花色品种也较少。除了颜色的不同外，蛋龙还可分为长尾和短尾两个不同的类型。在蛋龙的鉴赏中对背的要求是圆滑平顺，不能扛枪带刺；双眼凸出明显，并且一定要对称；游动灵活自如，静止时不倾头或趴缸。

The Dragon Eyes breed of the Eye Variation from Egg-Fish family, named Egg-Fish with Dragon Eyes, called Egg-Fish with Dragon Eyes in short. In the past, the quantity of the Egg-Fish with Dragon Eyes was not too much. Genetics experts had analyzed the genetic reasons for that why it is difficult to cultivate Egg-Fish with Dragon Eyes. But at present, with the continuous effort of goldfish industry, more and more Dragon Eyes or compound variation categories with Dragon Eyes in Egg-Fish family have occurred. Previously, the Dragon Eyes in Egg-Fish family was classified as a separate class, called Dragon Back. Together with Common Goldfish, Fantail Goldfish, Dragon Goldfish and Egg, to be the fifth category. While, this book does not take this method of classification. The Egg-Fish with Dragon Eyes is a small popularized variety and are not too many at present. And also the color varieties are not so many. Besides different colors, the Egg-Fish with Dragon Eyes can be also divided into two types, the long tails and the short tails. In the appreciation of the Egg-Fish with Dragon Eyes, the back should be smooth and flat, without agenesis blunt fin on the position of dorsal fin. The eyes convex obviously and must be symmetrical, When swimming, it should be flexible and comfortable and do not tilt the head or stay still on the cylinder when still.

红蛋龙
—— Red Egg-Fish with Dragon Eyes

- **花色类别** / Color
  红 / Red

- **品种类别** / Species
  蛋形 > 眼型变异类 > 龙睛型
  Egg-Fish > Eye Variation > Dragon Eyes

- **品种详述** / Species Description

　　红蛋龙与黑蛋龙系出同门，对外部形态的要求也一致。图例所示为红蛋龙，总体形态较为匀称。红蛋龙的体质强健，游动迅速，比较易于饲养。二龄以后的蛋龙眼睛特别大，对鱼进行操作时要小心，如果弄掉了眼睛就无法再复原，使其失去了观赏价值。蛋龙出现蒙眼现象（即透明的角膜出现混沌感）一般为细菌感染，及时换水并添加消炎药物，很快就能使其恢复。

Red Egg-Fish with Dragon Eyes and Black Egg-Fish with Dragon Eyes are from the same family. Their characteristics should be the same. As shown in the illustration is a Red Egg-Fish with Dragon Eyes. Its overall shape is symmetrical. Its physique is strong, moving quickly and easier to feed. The eyes of egg-dragon after 2 years stage are spectacular big. Be careful when operating on fish. If their eyes had been knocked off, they would not be able to recover so that they would lose ornamental value. Blindfolded phenomenon of Egg-Fish with Dragon Eyes (that is, the cornea appears transparent sense of chaos), in general, is resulted from bacterial infections. It can be recovery soon by changing the water and add anti-inflammatory drugs in time.

花色类别 / Color •

黑 / Black

品种类别 / Species •

蛋形 > 眼型变异类 > 龙睛型
Egg-Fish > Eye Variation > Dragon Eyes

品种详述 / Species Description •

黑蛋龙 — Black Egg-Fish with Dragon Eyes

　　黑蛋龙、黑龙睛和黑龙睛蝶尾如同三兄弟一般，总有割舍不掉的血缘感。图例所示为黑色长尾蛋龙，其全身漆黑如墨，眼球凸出眼眶之外，透明的角膜边缘有两道金色的虹彩，显得特别有神；背阔且厚，尾大且展；背部顺滑，两眼对称，是较优秀的一尾金鱼。黑蛋龙色彩目前还不太稳定，大多数会逐渐褪色为红蛋龙。

Black Egg-Fish with Dragon Eyes, Black Telescope and Black Butterfly Moor are like three brothers from the same family, they have a bond which can't be broken. The picture shows a Black Egg-Fish with Dragon Eyes Long Tail. Its whole body is pitch-black like ink, its eye pompons protrude outside the sockets and on the edge of its transparent cornea are two channels of golden rainbow. It has a very mystical appearance. The back is wide and thick, the tail is big and unfurled, the back is smooth, and the eyes are symmetrical. It's an outstanding goldfish. At the present the Black Egg-Fish with Dragon Eyes color hue is unstable; the majority of them will gradually fade into Red Egg-Fish with Dragon Eyes.

蛋形
Egg-Fish

∨

眼型变异类
Eye Variation

∨

蛤蟆眼型
Frog-Head

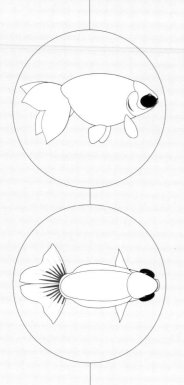

蛤蟆眼过去也叫蛤蟆头、蛙头，是一个古老的金鱼品种，也有把它划分在头型变异类中的。但目前金鱼鉴赏中，单独对蛤蟆头做玩赏的几乎没有，大多数情况是进行水泡的杂交育种时，将其作为一种中间的过渡体来使用，因此，将其归入眼型变异的范畴中较为妥当。在头型变异类的所有金鱼中，蛤蟆眼一般头茸不太发达，但在眼睛下方和眼眶之间，有一个硬质的小泡囊，很像青蛙的声囊，故而得名蛙头。过去资料中记载的蛤蟆眼花色品种比较多，后来逐步被各种花色的水泡所取代。

Frog-Head is an ancient goldfish breed. And it is sometimes be classified into the head variation category. As for the goldfish appreciation at present, little of them are supposed to be appreciated alone. In most cases, it is used as a transition body when cross-breeding Bubble-Eye. Therefore, it is appropriate to be classified into Eye Variation. Among all goldfish of head variation, the head pompon of Frog-Head is generally not well-developed, with a hard vesicle under the eyes and between eye sockets, which resembles frogs vocal sac. So it obtains its name frog head. In the past, many varieties of Frog-Head were recorded in the data, and then gradually replaced by various kinds of Bubble-Eye.

蓝蛤蟆眼 —— Blue Frog-Head

花色类别 / Color •

蓝 / Blue

品种类别 / Species •

蛋形 > 眼型变异类 > 蛤蟆眼型

Egg-Fish > Eye Variation > Frog-Head

品种详述 / Species Description •

　　图例所示为蓝蛤蟆眼，其头型扁方，嘴阔，头顶有鸡心形顶茸；眼球与眼眶之间有泡囊，内有淋巴液；背部平顺圆滑，尾鳍为大阔尾型；体形较修长，全身墨蓝色，鳞片边缘略带浅蓝色反光。蓝蛤蟆眼为小众品种，玩赏者不多，其已被蓝水泡所替代。是否应该恢复这个品种，目前业界尚有争议。

The picture shows a Blue Frog-Head. Its head is flat and square; its mouth is wide and the top of its head has a heart-shaped head pompon. Between the eye Pompons and eye socket is a vesicles filled with lymph. The back is smooth and slick, the caudal fin is a broad type. The body has a fairly slender figure and the whole body is dark blue with a faint reflective blue band on the edge of the scales. Blue Frog-Head is a small popularized variety and not many people appreciate it. And it has already been replaced by the Blue Bubble-Eye. There is a dispute in the goldfish industry over whether or not this breed should be rehabilitated and brought back to its former status.

雪青蛤蟆眼
Lilac Frog-Head

- **花色类别** / Color
  雪青 / Lilac

- **品种类别** / Species
  蛋形 > 眼型变异类 > 蛤蟆眼型
  Egg-Fish > Eye Variation > Frog-Head

- **品种详述** / Species Description

　　图例所示为雪青蛤蟆眼，由蓝蛤蟆眼杂交选育出的一种花色。雪青蛤蟆眼颜色较稳定，也属于小众品种，饲养极少。同大多数蛤蟆眼一样，是作为选育花色水泡的过渡体。不过雪青蛤蟆眼憨拙可爱，亦可作为一个奇特品种单独玩赏。其为杂交鱼，体质较强，比较容易饲养。

The legend shows Lilac Frog-Head, which is a variety selected and bred through hybridization of Blue Frog-Head. The colors of Lilac Frog-Head are relatively stable. It is also a minor variety and few are fed. Like most of the Frog-Head goldfish, it is also a transitional variety for selecting and breeding the fancy bubble. But Lilac Frog-Head is charmingly naive, so it can be appreciated alone as a peculiar variety. It is a hybrid goldfish and enjoys a good physique, so it is easier to feed.

雪青 / Lilac

红白蛤蟆眼
Red White
Frog-Head

花色类别 / Color •

红白 / Red White

品种类别 / Species •

蛋形 > 眼型变异类 > 蛤蟆眼型
Egg-Fish > Eye Variation > Frog-Head

品种详述 / Species Description •

　　图例所示为红白蛤蟆眼，是蛤蟆眼中较多的一类花色品种。红白蛤蟆眼的玩赏主要挑选巧色，如类似朱砂水泡的朱砂眼、全身洁白只有头顶呈现红色的朱顶、全身红色唯独头顶白色的玉顶等等。红白蛤蟆眼体质强健，易于饲养，可与许多其他品种混合饲养，多以俯视玩赏为主。

The picture depicts a Red White Frog-Head, which is a fairly common type among the Frog-Head. The pleasure of the Red White Frog-Head lies in color selection, like the similar Vermilion Frog-Head of Vermilion Bubble-Eye, which has a spotless white body with a red cap on the top of its head and the Red White Frog-Head has a spotless red body with a white cap on the top of its head. The Red White Frog-Head has a strong and healthy physique and is easy to rear and it is very suitable for rearing together with many other breeds. Its suitable for overlooking view.

蓝白蛤蟆眼
Blue White Frog-Head

- **花色类别** / Color
  蓝白 / Blue White

- **品种类别** / Species
  蛋形 > 眼型变异类 > 蛤蟆眼型
  Egg-Fish > Eye Variation > Frog-Head

- **品种详述** / Species Description

  图例所示为蓝白蛤蟆眼，蓝白花色又称羽衣，因为和黑白近似也称喜鹊花。黑喜鹊的黑白中，的确也夹杂一些蓝色，因此也不能说不对。蓝白色多为不稳定色，会逐渐褪白。蓝白蛤蟆眼属于小众品种，饲养玩赏的极少，也多作为杂交培养水泡花色的一个过渡体。蓝白水泡宜在温度低、水质硬的环境中玩赏，且以俯视为主。

  The picture shows a Blue White Frog-Head, the blue white color pattern is also known as Hagoromo. Because it is similar to black white, people also call it magpie pattern. A magpie bird does have some blue color mixed with its black and white, so the name magpie pattern can not be called wrong. The color is not stable and the white color will gradually fade away. The Blue White Frog-Head belongs to the small popularized breeds, very few people are interested in breeding it. And it is regarded as a transition form of crossbreeding the color pattern of Blue White Frog-Head. Blue White Bubble-Eye is suitable in low temperature, enjoying the hard water environment and mainly overlooking view.

花色类别 / Color •

红黑 / Red Black

品种类别 / Species •

蛋形 > 眼型变异类 > 蛤蟆眼型
Egg-Fish > Eye Variation > Frog-Head

品种详述 / Species Description •

红黑蛤蟆眼
—Red Black
Frog-Head

图例所示为红黑蛤蟆眼，其头型略像蛋形虎头，只是发育程度偏弱。眼球与眼眶之间有淋巴液囊；中短尾，背部较平顺，身形敦实，尾柄粗壮有力，游动能力不会太差；花色为红黑，应为黑色向红色转色的过渡色。短尾的红蛤蟆眼，已经被弓背短尾红水泡取代，玩赏的人不多。

The picture shows a Red Black Frog-Head and the shape of its head slightly resembles that of the Egg-Fish Lionhead, just not as prominently developed. Between the eye Pompons and eye socket is a lymph sac. It has a stocky body with a fairly smooth back and medium-size tail. The caudal body is stout strong, so its swimming ability is not too bad; It is red and black colored, and it should be transition color from black to red. The short-tailed Red Frog-Head has already been replaced by the Red Bubble-Eye with arched back and short tail. Its admirers are very few.

紫蓝蛤蟆眼
Chocolate Blue Frog-Head

- **花色类别** / Color

  紫蓝 / Chocolate Blue

- **品种类别** / Species

  蛋形 > 眼型变异类 > 蛤蟆眼型

  Egg-Fish > Eye Variation > Frog-Head

- **品种详述** / Species Description

  图例所示为紫蓝蛤蟆眼，中长尾，体形敦实，背部较平顺圆滑；头部头茸发育良好，类似蛋形金鱼中的蓝虎头；眼球与眼眶之间有淋巴液囊，泡囊大小适中，头阔嘴方；全身墨蓝色，鳞片边缘带淡蓝色银灰反光。如果能从这样的个体中选育出紫蓝鹅头水泡，那将是非常值得期待的事情。

  The picture shows a Chocolate Blue Frog-Head with a medium to long tail, a stocky figure and a smooth back angle; The head pompon is well-developed, similar to the Blue Lionhead (Tigerhead Type) of the Egg-Fish goldfish. Between the eye Pompons and eye socket is lymph sac, the size of the vesicle is moderate, its head is wide and its mouth is square. The whole body is dark blue. with shiny blue and silver bands on the edges of its scales. It would be something worth looking forward to, if Chocolate Blue Goosehead Bubble-Eye can be selective cultivated from this kind of individuals.

花蛤蟆眼
— Calico Frog-Head

花色类别 / Color •

软鳞五花 / Calico

品种类别 / Species •

蛋形 > 眼型变异类 > 蛤蟆眼型
Egg-Fish > Eye Variation > Frog-Head

品种详述 / Species Description •

　　图例所示为花蛤蟆眼，长尾，侧视体形和花蛋凤很像。其头茸发育不很明显，眼球与眼眶之间有淋巴液囊。蛤蟆眼体形不能太单薄，敦实一点会比较好，因此在蛤蟆眼的鱼苗期一定要给予充足的蛋白质饵料。花蛤蟆眼的花色略微显脏，不如素色的好看，如能选育出素色五花会更好。

The picture shows a Calico Frog-Head, it has a long tail and when viewed from the side it closely resembles the Calico Egg-Fish Phoenix Tail. The head pompon growth is not very evident. Between the eye Pompons and eye socket is lymph sac. The body of a Frog-Head should not be too thin and frail, and a stocky body is preferred, therefore it is important to provide feeding stuff that is rich in protein during the fry stage. The Calico Frog-Head has a color pattern which looks somewhat filthy and it's not as beautiful as a plain color. It is better to collective breeding pale Calico.

蛋形
Egg-Fish
∨
眼型变异类
Eye Variation
∨
望天型
Celestial-Eye

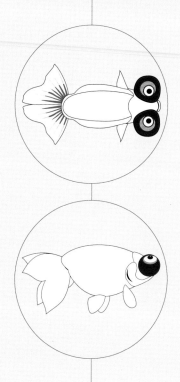

　　望天，又叫朝天龙、顶天眼，有记载约在1870年出现，是中国传统金鱼品种之一。因其两眼外凸并向上翻转，故而得名。望天两眼向上，有仰望天子的寓意，因此，它出现后，成为清朝宫廷中最得宠的金鱼品种之一。望天眼的瞳孔向上，俯视玩赏时，可以发现，眼球从外到内依次呈现银色、红色、金色三道金属质感光泽的环圈，俗称"三环套月"。能呈现三环套月的望天并不是很多，因此异常珍贵，是望天中的极品。说到望天，人们还会津津乐道一个传说，据说望天是因为养在一个黑暗的大水缸中，顶端只留一小孔透光和投放饲料，久而久之便形成了望天这个品种。当然，这只是人们一种美好的推想，虽是讹传，但却给予这个品种一种意趣。望天有长尾和短尾两种，北方饲养的略显短粗，而南方饲养的略显细长。

Celestial-Eye, also called Sky-Staring Dragon Eye, is one of Chinese traditional gold fish varieties. There was record that it occurred in 1870. Due to its two eyes convex and turn over, it obtains its name. Its two eyes turn upward morally, which means that it looks up the god which has the moral that it looks up the emperor(in ancient time of China, the emperor was regarded as the son of the heaven). Therefore, when it occurred, it became one of the most popular variety in the court in Qing dynasty. The pupils of Celestial-Eye are upward. When overlooking, from the outside to inside, the eye Pompons successively show the three rings in colors of silver, red, gold with metallic luster commonly known as "moon in the three rings". The Celestial-Eye which has the " moon in the three rings" is not too many, so it is particularly rare and is the master work in Celestial-Eye family. Speaking of Celestial-Eye, people will relish a legend:
It was said that the Celestial-Eye was kept in a black fish bowl, with only a orifice on the top for Light transmission and feeding. As time passes, it developed into Celestial-Eye variety. Of course, this is just a good guess of people. Although it is the rumor, but it brings a taste of this species. There are two types of Celestial-Eye, long tails and short tails. The tails of Celestial-Eye reared in northern is relatively more stubby, while in southern its tail is relatively long and thin.

红望天 — Red Celestial-Eye

花色类别 / Color •

红 / Red

品种类别 / Species •

蛋形 > 眼型变异类 > 望天型

Egg-Fish > Eye Variation > Celestial-Eye

品种详述 / Species Description •

　　图例所示为红望天，长尾类型。长尾红望天体态修长，颇具柔美之感，目前多在扬州等地饲养，具有悠久的历史，是当地著名的地方代表品种。由于二龄后观赏性不如一龄，因此，红望天大多只作当岁养殖并销售。红望天体质强健，较耐缺氧环境，易于饲养，目前在徐州、南阳等坑塘饲养地区较为常见。

The picture shows a Red Celestial-Eye of the long-tail type. The long-tailed Red Celestial-Eye has a slender body, soft and charming. At present it is mostly reared in Yangzhou, where it has a long history and is a famous local signature breed. The appreciation value in its second year stage is less than that in the first year stage, therefore, the majority of Red Celestial-Eye are just reared and sold in its first year stage. The Red Celestial-Eye has a strong and healthy physique, is able to endure living in an environment which has less than normal amounts of oxygen and is quite easy to rear. At present it is common kept in waterhole and pond breeding area in Xuzhou and Nanyang.

五花望天
Calico Celestial-Eye (Bluish Base)

- **花色类别** / Color
  软鳞五花 / Calico (Bluish Base)

- **品种类别** / Species
  蛋形 > 眼型变异类 > 望天型
  Egg-Fish > Eye Variation > Celestial-Eye

- **品种详述** / Species Description

图例所示为五花望天的亚成体，短尾类型。其背部平滑，尾柄粗壮；两眼外凸并呈90度上翻，头宽嘴阔；体色为软鳞五花，橙色色块略微显得偏多，是一个缺憾。目前五花望天数量不多，过去的图谱中多为文字记述而未见其图，可见，这一花色的望天品种还需慢慢筛选，有待逐步完善和成熟。

The picture shows a juvenile Calico Celestial-Eye (Bluish Base) of the short-tail type. Its caudal body is stout and strong; Two eyes are convex and turn over to be 90 degree. Both the head and mouth are wide. The color is Calico (Bluish Base) but what is regret is that the orange lump is a little bit more. At present there are not very many Calico Celestial-Eye (Bluish Base) in existence. In the past, the records about Calico Celestial-Eye (Bluish Base) in the illustration were most text without pictures, which means this type of Celestial-Eye variety should be selected step by step to make it more perfect and mature.

软鳞五花 / Calico (Bluish Base)

蛋形
Egg-Fish

ᐯ

眼型变异类
Eye Variation

ᐯ

水泡型
Bubble-Eye

水泡是中国传统金鱼重要代表品种之一，其历史非常悠久，可追溯到19世纪初期。水泡的外部形态与其始祖金鲫鱼变异跨度最大，两个丰满的泡囊如大红灯笼一般喜庆吉祥，游动时动感十足，是广受欢迎的名贵金鱼品种。蛋形金鱼中的水泡型分为两大类，即短尾弓背型和长尾平背型。其花色亦很丰富，目前可以看到的有红水泡、白水泡、墨水泡、蓝水泡、紫水泡、雪青水泡、红白水泡、紫白水泡、三色水泡、五花水泡等等。由于水泡不耐缺氧、饲养难度大、损耗高，许多金鱼养殖场目前都不愿饲养。在蛋形水泡的基础上，还会出现一些复合变异类，如下颌部出现颌泡的戏泡、眼睛角膜出现变异的灯泡眼水泡、鳞片发生变异的珍珠鳞水泡等等。

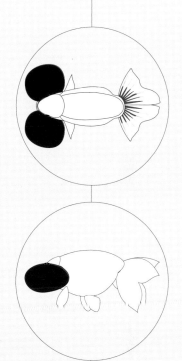

The Bubble-Eye is an important representative of China's traditional goldfish breeds. It has an extremely long history that can be traced back all the way to the early 19 century. Its characteristics variation is the most different from that of its ancient gold carps. Two plump vesicles are as happiness and auspicious as two red lanterns and when swimming it is full of dynamics. It is a very popular and famous goldfish breed. Bubble-Eye of Egg-Fish can be divided into two categories: short tail with arch-shaped back and long tail with flat back. Its color variations are plentiful, at present there are Red Bubble-Eye, White Bubble-Eye, Black Bubble-Eye, Blue Bubble-Eye, Chocolate Bubble-Eye, Lilac Bubble-Eye, Red White Bubble-Eye, Chocolate White Bubble-Eye, Red Black White Bubble-Eye, Calico Bubble-Eye and so on. Due to the Bubble-Eye being unable to endure living in an environment which has a low oxygen level it is difficult to rear, and many individuals are lost. Many goldfish breeders are now unwilling to raise it. The traditional place of Bubble-Eye production is in Jiangsu province, specifically the cities of Nanjing, Suzhou, and Xuzhou. On the basis of Egg-Fish Bubble-Eye, there are some compound variation, for example, Bubble-Eye with Two Bubble on Lower Jaw, the variation Bubble-Eye with Protruding Cornea on the cornea and pearlscale Bubble-Eye from scales variation.

红水泡

— Red Bubble-Eye

- **花色类别** / Color

  红 / Red

- **品种类别** / Species

  蛋形 > 眼型变异类 > 水泡型

  Egg-Fish > Eye Variation > Bubble-Eye

- **品种详述** / Species Description

  红水泡是最常见到的蛋形水泡，最著名的产地是江苏的扬州和徐州。红水泡的泡体特别发达，二龄以上的红水泡泡体可以大如鸡卵。水泡金鱼的泡囊内充满淋巴液体，所以分量较重，因此，在饲育水泡时，从幼苗期开始就要给予充足的生长空间和高蛋白质饵料，促使其身形粗壮饱满，有足够的力量可以平衡泡囊的重量。否则，因泡囊过重，二龄以后的水泡会趴缸，使其失去欣赏价值。

  The Red Bubble-Eye is the most common form of Egg-Fish Bubble-Eye and the most famous production areas are Yangzhou and Xuzhou in Jiangsu province. The Bubble-Eye's body of the Red Bubble-Eye is especially well-developed, at age two and up it can be as big as a chicken ovum. The Bubble-Eye vesicle of Bubble-Eye if full of lymph which makes it heavy, therefore, it is necessary to provide both ample room to grow and high protein feed at an early stage to ensure that it grows big and strong enough to support the weight of the Bubble-Eye vesicles. Otherwise, due to the too heavy Bubble-Eye vesicles, the Bubble-Eye after 2 years stage will just lie on the bottom of the cylinder and lose its appreciative value.

花色类别 / Color •

黑 / Black

品种类别 / Species •

蛋形 > 眼型变异类 > 水泡型
Egg-Fish > Eye Variation > Bubble-Eye

品种详述 / Species Description •

黑水泡
Black Bubble-Eye

　　黑水泡又称墨水泡，是较受欢迎的传统名贵金鱼品种之一。优秀的黑水泡全身拥有如同黑绒布般的质感。黑水泡有两种获得途径，一是在红水泡中选育出，二是在五花水泡中的青色个体可以选育出，而后者的黑色稳定性较好，并且时间愈长，色泽愈黑。图例所示为黑水泡金鱼，其腹部已有转色迹象。黑水泡性格活泼，与人互动性强，饲养时注意低密度，以防缺氧。

The Black Bubble-Eye also known as the Black Bubble-Eye is a fairly popular breed among the traditional well-known goldfish. The body texture of outstanding Black Bubble-Eye is as black velvet. There are two ways to get a Black Bubble-Eye, the first is by selectively breeding Red Bubble-Eye; the second is by selectively breeding a Lilac individuals of Calico Bubble-Eye, while the black color of the latter is more stable. And the longer the time, the darker the color. The picture shows a Black Bubble-Eye, its abdomen already shows signs of color change. The Black Bubble-Eye has a lively nature and interacts with human beings. It should be low density when breeding, in order to prevent hypoxia.

蓝水泡 — Blue Bubble-Eye

- **花色类别** / Color
  蓝 / Blue

- **品种类别** / Species
  蛋形 > 眼型变异类 > 水泡型
  Egg-Fish > Eye Variation > Bubble-Eye

- **品种详述** / Species Description

　　蓝水泡过去较稀有，目前发展很快。蓝水泡分为两大类，即短尾弓背型和长尾平背型。图例1为短尾弓背型蓝水泡，是近年来金鱼业者利用红弓背水泡和蓝平背水泡杂交选育、逐步提纯，获得稳定遗传的一个新品种。其特点是身体粗壮，泡囊适中，二三龄后也不会趴缸，依然活力十足。图例2是长尾平背型蓝水泡。

Blue Bubble-Eye is quite rare but the numbers have been growing rapidly at present. The Blue Bubble-Eye are divided into two major types: short tail, arch back and long tail, flat back. Picture 1 shows a Blue Bubble-Eye with short tail and arch back which is the result of selective cross breeding Red Bubble-Eye with arch back and Blue Bubble-Eye with arch back and refining step by step, to obtain a new breed with stable heredity by goldfish industry in recent years. Its characteristic feature is the sturdy body, the size of the Bubble-Eye vesicle is moderate. It will not lie still on the cylinder and will also full of spirit even after two or three years stage. Picture 2 shows a long tail, flat back Blue Bubble-Eye .

花色类别 / Color •

紫 / Chocolate

品种类别 / Species •

蛋形 > 眼型变异类 > 水泡型

Egg-Fish > Eye Variation > Bubble-Eye

品种详述 / Species Description •

紫水泡 — Chocolate Bubble-Eye

　　紫水泡近些年来较少出现，是水泡中比较珍稀的花色品种。紫水泡出现在20世纪50年代到60年代，据资料介绍是从红水泡中选育的，其色彩稳定性不强，不少还是会褪色成红水泡。紫水泡的色彩表现受水质影响也较大，一般在水质较硬或偏弱碱性水中呈现赤铜色，而在水质较软或微酸性水体中会呈现黄铜色甚至青铜色。

In recent years the Chocolate Bubble-Eye has seldom been seen. It's a relatively rare variety of Bubble-Eye. It first appeared sometime between 1950 and 1960 and according to sources, it was the result of selectively breeding Red Bubble-Eye. Its color stability is weak and as a result many individuals fade and become Red Bubble-Eye. The color of the Chocolate Bubble-Eye is to a large extent influenced by water quality. Generally, in harder or weak alkaline water it shows red copper color, and in soft or slightly acidic water it will appear in brass or even bronze.

红白水泡

Red White Bubble-Eye

- **花色类别** / Color
  红白 / Red White

- **品种类别** / Species
  蛋形 > 眼型变异类 > 水泡型
  Egg-Fish > Eye Variation > Bubble-Eye

- **品种详述** / Species Description

　　红白水泡是最受欢迎的水泡花色品种，其俏皮的外形和艳丽的花色，时常让观赏者长久驻足，不忍离去。红白水泡中，最知名的便是朱砂水泡了。朱砂水泡全身洁白，唯独两个泡囊通体呈朱砂红色，其出现概率极低，是许多水泡爱好者梦寐以求的无上至宝。朱砂水泡要求整个泡囊都是红色，这个品相较难达到，常见的是泡体上半部呈红色的，称作朝阳泡。红白水泡的泡囊大而薄，饲养时要极其小心，防止弄破。泡囊破损后，如伤口不大，大多可自愈。如果伤口过大，则很难恢复。有时水泡的两个泡囊发育会逐步不对称，可以通过一些人为的手法加以调节，达到最佳观赏效果。

The Red White Bubble-Eye is the most popular variety of Bubble-Eye. Its delightful shape and gorgeous color pattern often stops people in their tracks. Among Red White Bubble-Eye, the most famous is Vermilion Bubble-Eye and the Vermilion Bubble-Eye has a spotless white body, only the two Bubble-Eye vesicles assume a cinnabar red color which is extremely rare and many Bubble-Eye enthusiasts dream of obtaining such a treasure. Vermilion Bubble-Eye requires the whole vesicle is red, which is quite difficult to achieve. The common one is like that, the upper part is red, called Rising Sun Bubble. The vesicle of Red White Bubble-Eye is big and thin and one must be extremely careful to prevent injuries. If the vesicle is injured but the wound is not too large, it will usually heal on its own. If the wound is large it is very difficult to restore it to its former glory. Sometimes the two vesicles of the Bubble-Eye grow asymmetrically. By some artificial manners to adjust breeding to make it get the most appreciative value.

紫白水泡

——Chocolate White
Bubble-Eye

- **花色类别** / Color
  紫白 / Chocolate White

- **品种类别** / Species
  蛋形 > 眼型变异类 > 水泡型
  Egg-Fish > Eye Variation > Bubble-Eye

- **品种详述** / Species Description

　　紫白水泡是比较罕见的水泡花色品种，原为苏州的金鱼业者从三色水泡中选育出，但因为遗传不稳定等一些因素，后来逐渐遗失。图例所示紫白水泡，由紫水泡选育而出，其遗传并不稳定。但随着金鱼业者的不断努力，通过几年时间的提纯，应该可以获得稳定遗传的个体。紫白水泡体质较弱，对水质要求相对较高。

The Chocolate White Bubble-Eye is a type of Bubble-Eye which can seldom be seen. It is originally selective cultivated from Red Black White Bubble-Eye by goldfish industry in Suzhou. But because of unstable heredity and other factors, it was later gradually lost. The picture shows a Chocolate White Bubble-Eye, selective cultivated from Chocolate Bubble-Eye and its heredity is not at all stable. But with great efforts and years of refinement, the goldfish industry should be able to obtain an individual with stable heredity. The Chocolate White Bubble-Eye has a rather weak physique and demands great water quality.

五花水泡
—Calico Bubble-Eye
(Bluish Base)

花色类别 / Color •

软鳞五花 / Calico (Bluish Base)

品种类别 / Species •

蛋形 > 眼型变异类 > 水泡型
Egg-Fish > Eye Variation > Bubble-Eye

品种详述 / Species Description •

　　五花水泡有两种风格：一种是以苏州为代表的，如图例所示的五花水泡，其特点是泡圆、身长、尾长、尾叶窄而尖；另一种风格是以西安为代表的五花水泡，其特点是泡椭圆（当地俗称腰子泡）、身短、尾短、尾叶略宽、先端较圆，该种水泡目前已很罕见。图例所示的两尾五花水泡无论是形体还是花色，都近乎完美，较具代表性。五花水泡对水质要求较严，如水质恶化，其泡囊中的淋巴液会变成乳白色，继而发炎变红，如不及时调节水质，很容易造成损失。

There are two different styles of Calico Bubble-Eye. The first is Suzhou style, as shown in the picture. its distinguishing features are like that: the Bubble-Eye is round, the body is long, the tail is long, and the tail leaf is narrow and pointed. The other style of Calico Bubble-Eye's representative is Xi'an and its distinguishing features are like that, the Bubble-Eye is oval (locally known as kidney Bubble-Eye), the body is short, the tail is short, the tail leaf is a slightly wide and the first part is round. Xi'an style has been seldom seen in nowadays. The two Calico Bubble-Eye shown in the illustration, no matter its body shape or its color are almost perfect and representative. The Calico Bubble-Eye needs very high water quality, otherwise the lymph in its vesicle turns milky white then red due to inflammation. If the water quality is not promptly adjusted it can easily lead to a loss.

蛋形
Egg-Fish

∨

鼻膜变异类
Nasal Variation

∨

蛋凤球型
Egg-Fish Phoenix Tail with Pompons

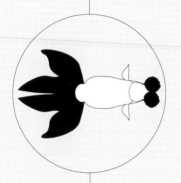

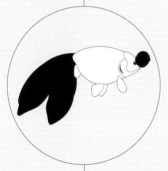

　　蛋凤球是近些年来，由蛋球和蛋凤杂交选育出的一个品种。其融合了蛋凤和蛋球的特点，发展成俯视和侧视效果俱佳的一个品种类型。蛋凤球在外部形态特征上，总体保留了蛋凤修长飘逸的特点，而在欣赏点较少的头部，点缀上两个绒球。蛋凤的审美在于简约朴实，而加上两个绒球既丰富了看点，又不会对原来的审美造成破坏。蛋凤球选育的时间不长，花色也不多，常见的有红色、红白色、青色和五花色。蛋凤球体质较强，耐缺氧，是相对易于饲养的一个金鱼品种。蛋凤球也属于小众的金鱼品种，饲养者不多，目前出产于福州、南京、淮安等地。蛋凤球在外部特征上要求背部平顺，尾鳍修长为上，不打折翻卷，绒球作为点缀不宜过大，适中最好。

Egg-Fish Phoenix Tail with Pompons variety is the result of cross breeding Egg-Fish with Pompons and Egg-Fish Phoenix Tail. Its characteristics mix that of the Egg-Fish Phoenix Tail and the Egg-Fish with Pompons. It has developed into a breed which is suitable for both overlooking and side looking. The characteristics of Egg-Fish Phoenix Tail with Pompons, as a whole, carry on Egg-Fish Phoenix Tail's characters of slender and elegant and there are two Pompons on its head on which there is little appreciative value. The charm of the Egg-Fish Phoenix Tail lies simple and plain and plus two Pompons can not only enrich its charm but not damage the original aesthetic. Egg-Fish Phoenix Tail with Pompons' breeding time is not long, and the color varieties are not too many. Its common color varieties are: red, red and white, blue and calico. The Egg-Fish Phoenix Tail with Pompons' physique is strong and it is tolerance to hypoxia. It is a gold ship breed that relatively easy to rear. Egg-Fish Phoenix Tail with Pompons is also a small popularized variety and has not so many breeders. At present it is produced in Fuzhou, Nanjing, Huai'an area. The characteristics of Egg-Fish Phoenix Tail with Pompons requires that the black is flat and smooth, caudal fin is narrow and long, not turning over and the Pompons should not be too big, moderate is enough.

花色类别 / Color •

红 / Red

品种类别 / Species •

蛋形 ＞ 鼻膜变异类 ＞ 蛋凤球型

Egg-Fish ＞ Nasal Variation ＞ Egg-Fish Phoenix Tail with Pompons

品种详述 / Species Description •

红蛋凤球
—— Red Egg-Fish Phoenix Tail with Pompons

　　图例所示为红蛋凤球，当岁鱼，其外部形态各部分比例均比较协调，绒球大小也适中，但背部的平整度略显缺憾。其尾鳍与身体的比例基本接近1：1，二龄以后尾鳍还会加长，并会自然下垂，呈现出凤尾的独特韵味。其腹部鳞片略有松散，是当岁生长速度过快、鳞片上钙质沉积速度跟不上所致。因此，幼鱼期控制好生长节奏，对鳞片的发育非常重要。

As shown in the illustration is Red Egg-Fish Phoenix Tail with Pompons in the first year stage. Each part proportion of its body is coordinate. The pompon is in moderate size but the back is uneven. The proportion of the caudal fin and body is almost 1:1. After 2 years stage, the caudal fin will grow longer and be naturally drooping, showing the unique charm of phoenix tail. Its scales on the belly are relatively loose and scattering, because it grows so rapid that the scale calcium position rate can not keep up with the growth speed. Therefore, it is very important for the scale development to control the growth pace during young fish period.

红白蛋凤球

Red White Egg-Fish
Phoenix Tail with Pompons

- **花色类别** / Color

  红白 / Red White

- **品种类别** / Species

  蛋形 > 鼻膜变异类 > 蛋凤球型

  Egg-Fish > Nasal Variation > Egg-Fish Phoenix Tail with Pompons

- **品种详述** / Species Description

　　图例所示为红白蛋凤球，其整体发育良好，比例协调，尾巴宽大舒展，朱眼赤球，是比较优秀的一尾蛋凤球。20世纪20到30年代，天津的著名工笔走兽翎毛画大师刘奎琳先生画过不少以金鱼为题材的中国画，其中就有好几幅是表现红白蛋凤球的。例如一幅扇面作品，描绘的是一对全身洁白，唯独头顶一块红斑，并带有两个朱砂球的蛋凤球，神态惟妙惟肖，令人心动。

As shown in the illustration is Red White Egg-Fish Phoenix Tail with Pompons. Its overall development is good. Each part proportion of its body is coordinated well. It is an excellent Egg-Fish Phoenix Tail with Pompons with a widely stretching tail, red eyes and body. In the 20th century, 20 or 30 years, Tianjin well-known beast feather painting master Mr. Liu Kuilin painted lots of Chinese painting with theme of goldfish among which there are several paintings are Red White Egg-Fish Phoenix Tail with Pompons. For example, there is a fan works, depicting a pair of Egg-Fish Phoenix Tail with Pompons which are totally white except for a red stain on the head and have two big cinnabar Pompons. Very exciting and vivid.

五花蛋凤球

Calico Egg-Fish Phoenix Tail with Pompons (Bluish Base)

- **花色类别** / Color

  软鳞五花 / Calico (Bluish Base)

- **品种类别** / Species

  蛋形 > 鼻膜变异类 > 蛋凤球型

  Egg-Fish > Nasal Variation > Egg-Fish Phoenix Tail with Pompons

- **品种详述** / Species Description

  图例所示为当岁的五花蛋凤球幼鱼，其整体形态比例均较为协调，花色比较素雅。和大多数长尾金鱼一样，其尾先有些翻折，使观赏性要打折扣。五花蛋凤球体质强健，易于饲养，而软鳞的特征，又是杂交品种的特性，这可以使它成长为身形比较大的金鱼。五花蛋凤球和人的亲和度好，互动性高，遗传稳定，是近几年来出现的较优秀的新品种之一。

  As shown in the illustration is the young fish of Calico Egg-Fish Phoenix Tail with Pompons (Bluish Base). Its body is in proportional coordination. The color is relatively pale. Like most of the long tail goldfish, some part of its tail edge is turnover. As a result, the appreciative value is reduced. Calico Egg-Fish Phoenix Tail with Pompons (Bluish Base)'s physique is vigorous, and it is easy to rear. It can grow in big size because it has the characteristics of both matt type and hybrid breeding. Calico Egg-Fish Phoenix Tail with Pompons (Bluish Base) has good affinity and high interaction and it is genetically stable and is one of new excellent breeds.

蛋形
Egg-Fish

∨

鼻膜变异类
Nasal Variation

∨

蛋球型
Egg-Fish with Pompons

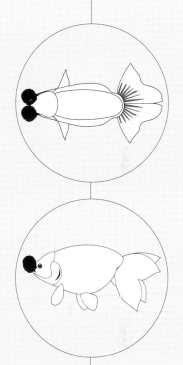

　　蛋球是蛋形金鱼里较古老的一个品种，从出现到现在，已经有超过100年的历史了。蛋球由蛋鱼演化而来，其外形特征也基本和蛋鱼一样。其身体短圆，外形似蛋，头尖，由发达的鼻瓣膜形成的绒球点缀在吻部上方，一般双球的较多见，而四球的则十分罕见。绒球类的金鱼在日本被称为花房，意思是鼻房（腔）如同花状，而中国人则赋予它一个更具诗意的名称——簪花。蛋球以前多产于京津一代，花色品种也很多，有红蛋球、银蛋球、黑蛋球、蓝蛋球、紫蛋球、红白蛋球、五花蛋球等等，但随着该品种的小众化，很多花色已经湮灭，常见的只有五花蛋球和红蛋球。现在通过金鱼业者的努力，已经恢复了紫蛋球和红白蛋球。

Egg-Pompons is an ancient breed. From its occurrence up till now, it has over 100 years of history. Egg-Pompons developed from Egg-Fish. Its characteristics are almost like that of Egg-Fish. Its body is short and round, like an egg with a pointed head. The developed nasal valve formed into Pompons embellished above the snout. Generally, the double Pompons are common while the four Pompons are very rare. The Pompons in Japan is known as flower house, which means that nasal cavity is in flower shape. While Chinese people give it a poetic name: "Zan Hua" (a kind of flower ornament wear on head). In the past, most egg-Pompons fish were reared in Beijing and Tianjin area. There were a lot of categories of Egg-Fish, such as Red Egg-Fish with Pompons, Silver Egg-Fish with Pompons, Black Egg-Fish with Pompons, Blue Egg-Fish with Pompons, Chocolate Egg-Fish with Pompons, Red White Egg-Fish with Pompons and Calico Egg-Fish with Pompons (Bluish Base) and so on. With the de-commoditizing of the variety, many color varieties have been extinct. Only Calico Egg-Fish with Pompons (Bluish Base) and Red Egg-Fish with Pompons are common. Through hard work of goldfish industry, Chocolate Egg-Fish with Pompons and Red White Egg-Fish with Pompons have reappeared.

紫蛋球 | Chocolate Egg-Fish with Pompons

- **花色类别** / Color
  紫 / Chocolate

- **品种类别** / Species
  蛋形 > 鼻膜变异类 > 蛋球型
  Egg-Fish > Nasal Variation > Egg-Fish with Pompons

- **品种详述** / Species Description

　　图例所示为紫蛋球。紫蛋球是传统品种，但后来灭绝了。现在看到的紫蛋球，是近几年来，金鱼业者以五花蛋球进行杂交繁育而成的。较之传统的五花蛋球的尖头，紫蛋球头方嘴阔，全身紫金色，背部圆顺平滑，尾柄较粗壮，游动时矫健有力。紫蛋球有部分会褪色成红色蛋球，偶尔也会出现全身紫色，唯独两个绒球呈红色的"紫裳红妆"，十分漂亮。

As shown in the illustration is Chocolate Egg-Fish with Pompons, which is a traditional variety. Later it has gone extinct. Chocolate Egg-Fish with Pompons at present is the result of hybrid breeding Calico Egg-Fish with Pompons (Bluish Base) by goldfish industry in recent years. Compare with the pointed head of traditional Calico Egg-Fish with Pompons (Bluish Base), Chocolate Egg-Fish with Pompons has a square head, a wide mouth, and a totally gold-Chocolate body. Its back is round and smooth and the caudal body is strong which is vigorous when swimming. Some Chocolate egg Pompons can fade into red Egg-Fish with Pompons, occasionally totally Chocolate. Only the one who has two red Pompons, "Chocolate dress red make up", is pretty beautiful.

樱花蛋球
—Sakura Egg-Fish with Pompons (Red White Matt)

花色类别 / Color •

软鳞红白 / Red White Matt

品种类别 / Species •

蛋形 > 鼻膜变异类 > 蛋球型
Egg-Fish > Nasal Variation > Egg-Fish with Pompons

品种详述 / Species Description •

　　图例所示为樱花蛋球。樱花蛋球是五花蛋球中出现的软鳞红白色的个体，过去对软鳞的红白色金鱼和硬鳞的红白色金鱼没有区分，统称为红白。当日本金鱼进入国内后，逐渐将软鳞红白色的名称"樱花"推广到中国金鱼的命名上来。所以，我们看到在过去的图谱中，樱花蛋球的名称都是红白蛋球。樱花蛋球以白多红少的白胜为贵，红色部分要少，但颜色最好呈现殷红色，如巴林的鸡血冻石一般。

As shown in the illustration is Sakura Egg-Fish with Pompons (Red White Matt) which are the Red White Matt individuals from Calico Egg-Fish with Pompons (Bluish Base). In the past there was no classification about matt and hard-scale Red White goldfish, both of them are called Red White. After Japanese goldfish introduced into China, the name of Red White Matt: "Sakura" has gradually popularized into Chinese goldfish's name. So, in the past, Sakura Egg-Fish with Pompons (Red White Matt) was named Red White Egg-Fish with Pompons in the illustration. As for the Sakura Egg-Fish with Pompons (Red White Matt), more white is better; the red part should be less but the color should be in blackish red, as Balin Chicken Blood Stone.

虎纹蛋球
——Tiger Banded Egg-Fish with Pompons

- **花色类别** / Color
  软鳞红黑 / Tiger Banded

- **品种类别** / Species
  蛋形 > 鼻膜变异类 > 蛋球型
  Egg-Fish > Nasal Variation > Egg-Fish with Pompons

- **品种详述** / Species Description

  图例所示为虎纹蛋球，俗称虎皮花蛋球。其全身橙黄色，并缀以黑色条纹状花斑，如同猛虎身上的花纹。其头部橙红色，并有两个粉红色的绒球，如同两朵盛开的牡丹。

  As shown in the illustration is Tiger Banded Egg-Fish with Pompons, commonly known as Tiger's Skin Calico Egg-Fish with Pompons. Its body is totally orange-yellow, decorated with black striped piebald, as the tiger's body pattern. Its head is orange-red, with two pink pompons as two blooming peony.

花色类别 / Color •

软鳞五花 / Calico

品种类别 / Species •

蛋形 > 鼻膜变异类 > 蛋球型
Egg-Fish > Nasal Variation > Egg-Fish with Pompons

品种详述 / Species Description •

花蛋球
—— Calico Egg-Fish with Pompons

图例所示为花蛋球。花蛋球是花色介于五花和虎纹之间的蛋球。花蛋球是金鱼古老品种之一，其背型多直背，尾柄处结合较生硬，性格活泼好动，体质强健，较耐缺氧，易于饲养。花蛋球的绒球部分发达，需要适当修剪，使其不要松散。球体部分虽然可以再生，但操作时仍需多加小心。

As shown in the illustration is Calico Egg-Fish with Pompons. Its color pattern is a mix between Calico (Bluish Base) and Tiger Banded. Calico Egg-Fish with Pompons is one of ancient goldfish breed. Most of them have straight back. The combination of caudal body is rigid. It is active, vigorous, and tolerant to hypoxia. It is easy to rear. Part of the Pompons are developed which need to be proper trimmed to prevent it from scattering. Although Pompons part can be regenerated, people also should be careful when in operation.

花色类别 / Color •

软鳞五花 / Calico (Bluish Base)

品种类别 / Species •

蛋形 > 鼻膜变异类 > 蛋球型
Egg-Fish > Nasal Variation > Egg-Fish with Pompons

品种详述 / Species Description •

五花蛋球 —— Calico Egg-Fish with Pompons (Bluish Base)

　　五花蛋球又称五彩蛋球，是金鱼古老的品种之一，由于较为小众，五花蛋球一度濒临灭绝。五花蛋球过去多产于京津一代，现在全国各地均有饲养。五花蛋球以素兰花色配以双红球为贵，其次是素兰花色配以一红一白或一红一黑的鸳鸯球花色为佳。图例2为五彩花色配鸳鸯球，也十分漂亮。二龄以上的蛋球，头顶上会有薄薄的一层顶茸，如同一顶小方帽。五花蛋球的饲养应注意保持水质清新，其五彩的体色才会呈现出最好的状态。

Calico Egg-Fish with Pompons (Bluish Base), also named multicolored Egg-Fish with Pompons, is one of ancient variety of goldfish. Because it is only reared by small crowd, it is once on the verge of extinction. In the past, Most Calico Egg-Fish with Pompons (Bluish Base) are reared in Beijing and Tianjin area. But at present it is reared all over the country. Pale color with double red Pompons is the best, followed by pale blue color with two Pompons in different colors: red and white or red and black. As shown in the illustration 2 is calico color with two Pompons in different colors, is pretty beautiful. The Egg-Fish with Pompons in over 2 years stage has a very thin head pompon on its head, as a square hat. Pay attention to keeping water fresh, so that its colorful body can be in the best condition.

蛋形
Egg-Fish

∨

复合变异类
Compound Variation

∨

灯泡眼蛋龙型
Dragon-Back with Protruding Cornea

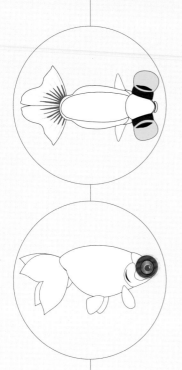

　　灯泡眼蛋龙是蛋形金鱼龙睛型中的一个突变个体，目前遗传尚不稳定。所谓灯泡眼，是指眼睛透明角膜外凸，如同白炽灯泡的玻璃灯头一样。灯泡眼多出于龙睛，而普通的平眼，也有出灯泡眼的变异个体。

Bulb-eye egg dragon is a variation individual of dragon-eyes. At present it is genetically unstable. Bulb-eyes means transparent eyes with convex corneal, like a glass lamp of the incandescent light. Most bugle-eyes is from dragon-eyes, while the buble-eyes variation individual can be also from common flat eyes.

花色类别 / Color •

红 / Red

品种类别 / Species •

蛋形 > 复合变异类 > 灯泡眼蛋龙型

Egg-Fish > Compound Variation > Dragon-Back with Protruding Cornea

品种详述 / Species Description •

红灯泡眼蛋龙——Red Dragon-Back with Protruding Cornea

图例所示为红灯泡眼蛋龙当岁鱼，灯泡眼型在一龄时发育不明显，呈现为劣弧，而在二龄以后则发育得很好，呈优弧状。操作灯泡眼时要小心，一旦破损无法修复，就会失去了观赏价值。

As shown in the illustration is Red Dragon-Back with Protruding Cornea in one year stage. The development of Protruding Cornea is not obvious in its first year stage, and the body is inferior arc. After two years stage, it will well-developed in superior arc. Operation should be careful. Once broken, it can not be recovery, and it will lose its appreciative value.

蛋形
Egg-Fish

ⅴ

复合变异类
Compound Variation

ⅴ

蛋龙球型
Egg-Fish with Dragon Eyes and Pompons

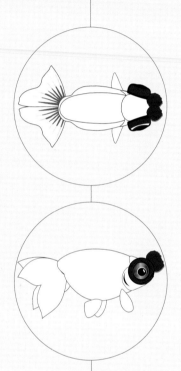

一般来说，蛋龙球由蛋龙发展而来。绒球型属于显性遗传，较容易获得，且遗传相对稳定。而蛋形金鱼中的龙睛较难稳定。笔者也曾在文形金鱼龙睛球的子代中，偶尔发现蛋龙球的突变个体，但繁殖的后代还是文形龙睛球。

Generally, Egg-Fish with Dragon Eyes and Pompons is developed from Egg-Fish with Dragon Eyes. Pompons are dominantly inherited, relatively easy to obtain, and relatively stable. While the Dragon Eyes in Egg-Fish is unstable. The author have by chance found the individual of egg-dragon variation in Dragon Eyes with pompons offspring of Fantail Goldfish, and he has occasionally found the variation individual of Egg-Fish with Dragon Eyes and Pompons of which the breeding offspring are still Fantail Goldfish.

花色类别 / Color •

红白 / Red White

品种类别 / Species •

蛋形 ＞ 复合变异类 ＞ 蛋龙球型
Egg-Fish ＞ Compound Variation ＞ Egg-Fish with Dragon Eyes and Pompons

品种详述 / Species Description •

红白蛋龙球
Red White Egg-Fish with
Dragon Eyes and Pompons

图例所示为红白蛋龙球当岁鱼，其背部圆滑平顺，绒球紧实对称，鱼鳍舒展，如能在二龄三龄以后变为凤尾，则会更加富有韵味。

As shown in the illustration is Red White Egg-Fish with Dragon Eyes and Pompons in one year stage. Its back is round and smooth, Pompons are tight and symmetric and fins are stretching. If it can turns into Phoenix Tail after 2 or 3 years stage, it will be more charming.

蛋形
Egg-Fish

∨

复合变异类
Compound Variation

∨

虎头球型
Lionhead with Pompons

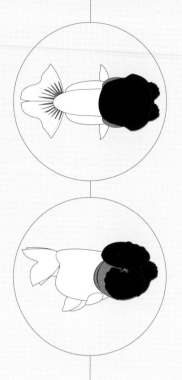

　　虎头球是蛋形金鱼虎头型与绒球型叠加的复合变异类。在蛋形金鱼中，许多狮头型会有较明显的鼻膜发育现象，有的甚至会发育为较好的绒球，古谱中将之妙称为"狮子滚绣球"。这种变异的形成，可能是因为以前在培育各种蛋形金鱼虎头型或狮头型时，金鱼业者为了快速获得丰富的色彩变异，而采用了将各种蛋球与之杂交的方式，因此会出现带球的现象。例如现在看到的蛋形金鱼蓝虎头就是有蓝蛋球的血缘，而五花虎头则有五花蛋球的血缘。而后，当各种花色的蛋形金鱼虎头型和狮头型开始逐渐趋向稳定，人们为了追求其头部的发育，而逐渐开始淘汰其带球的个体，因此，这些蛋形的虎头球和狮头球越来越少见。本书收录的蛋形虎头球只有一例，足见其珍稀。若金鱼从业者有心，也可以对这个品系加以关注，逐渐丰富完善。

Lionhead with Pompons is the compound variation of the superposition of Tigerhead and Pompons in Egg-Fish family. Among Egg-Fish, lots of Lionhead's nasal valve are developed obviously and some even develop into good Pompons, which has a wonderful name "the lion rolling Pompons". The reason probably is that in the past when cultivating various Tigerhead or Lionhead of Egg-Fish, the goldfish industry hybrid cultivate it with various Egg-Fish with Pompons in order to quickly obtain various color-variation, therefore, there are Pompons. For example, the Egg-Fish Blue Lionhead (Lionhead Type) we can see at present has blood relationship with Blue Egg-Fish with Pompons, while Calico Lionhead (Lionhead Type) (Bluish Base) has blood relationship with Calico Egg-Fish with Pompons (Bluish Base). Later, when various Tigerhead and Lionhead of Egg-Fish gradually grow stable, people are in pursuit of the head development, and gradually eliminate the individuals with Pompons so that the lionhead with Pompons (Tigerhead Type) and Lionhead with Pompons (Lionhead Type) of Egg-Fish are more and more rare. This book only contains one case of Egg-Fish Lionhead with Pompons (Tigerhead Type), which is enough to show how rare it is. If the goldfish industry cares about it, they can also be concerned about this variety and gradually enrich and improve it.

紫虎头球——Chocolate Lionhead with Pompons

花色类别 / Color •

紫 / Chocolate

品种类别 / Species •

蛋形 > 复合变异类 > 虎头球型

Egg-Fish > Compound Variation > Lionhead with Pompons

品种详述 / Species Description •

　　图例所示为紫虎头球，是蛋形金鱼虎头型与绒球型叠加的复合变异类。其头茸紧实，身形壮硕，嘴上部有两个发育完好的绒球，全身呈紫铜色。蛋形金鱼的头型变异类中，狮头型较虎头型要多，因此，蛋形金鱼的虎头球，要比蛋形金鱼的狮头球更加珍贵。虎头球是复合变异类，其两个变异位置都在头部，且发育程度适中，二者相映成趣。

As shown in the illustration is Chocolate Lionhead with Pompons, which is the compound variation of the superposition of Tigerhead and Pompons in Egg-Fish family. Its head pompon is tight and body is strong; There are well-developed Pompons above its mouth, and its body is totally chocolate copper. Among the Head Variation of Egg-Fish, Tigerhead is less than Lionhead, therefore, the Lionhead with Pompons (Tigerhead Type) of Egg-Fish is more rare. Lionhead with Pompons is compound variation, two variant positions are located on the head, and is moderately developed, which are side by side.

蛋形
Egg-Fish
∨
复合变异类
Compound Variation
∨
龙睛寿型
Ranchu with Dragon Eyes

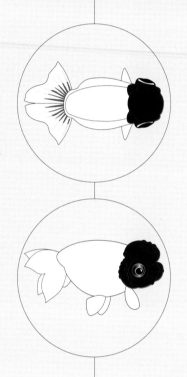

　　龙睛寿是福寿中出现的龙睛型的变异，是福州金鱼业者近几年创新的一个花色类型。其花色品种有红色、黑色、蓝色、黑白、三色和五花。由于是刚刚推出的新品，其市场接受度还有待提高。与龙睛寿相类似的品种有早几年河北金鱼业者培育的龙猫，即蛋形金鱼狮头型中出现的龙睛型，花色也较为丰富，当岁即发育良好，但二龄以后较易出现倾头的现象，因此比较侧重于当岁鱼的玩赏。而当岁鱼身形娇小，不能达到虎头型或狮头型的雄健，故而称之为龙睛猫狮或龙猫。龙睛寿应当控制好生长节奏，延长其观赏周期。

Ranchu with Dragon Eyes is the Dragon Eyes variation in Fuzhou Ranchu, which is a new color of variety, a result of an innovation in recent years by Fuzhou goldfish industry. The varieties has six color of types: red, black, blue, black and white, tri-color and calico. As is a new variety, its market acceptance has yet to be improved. Similar to Ranchu with Dragon Eyes, dragon-cat is cultivated a few years earlier by Hebei goldfish industry, which is Dragon Eyes of Lionhead in Egg-Fish family. It is rich in color, and has been well-developed when it is in one year stage. But its head is prone to tilt after 2 years stage. So it is the best time to view it when it is in 1 year stage. Because it is petite and dainty when it is 1 year stage and can not be as strong as Tigerhead or Lionhead, it obtain its name: Dragon-Cat. The growth pace should be well-controlled to extend its viewing period.

花色类别 / Color •

黑 / Black

品种类别 / Species •

蛋形 > 复合变异类 > 龙睛寿型
Egg-Fish > Compound Variation > Ranchu with Dragon Eyes

品种详述 / Species Description •

黑龙晴寿
——
Black Ranchu
with Dragon Eyes

　　图例所示为黑龙晴寿。其外部形态特征与黑兰寿相同，唯独眼睛呈现龙睛型。头茸发育相对滞后，如果头茸过早发育成熟，其倾头的比例会很高。龙睛寿背幅较宽，尾柄粗壮，背部圆滑，呈木梳背形或长抛物线背形，这样背形的鱼，后期会有足够的发育空间，成长为大鱼。如果龙晴寿体格不够大，则在生理上以及视觉上，无法平衡前半部的特点。

As shown in the illustration is Black Ranchu with Dragon Eyes. Its characteristics are the same with Black Ranchu with exception for its eyes of Dragon Eyes type. The development of head pompon relatively lag behind. If its head velvet prematurely well-developed, its head is extremely prone to tilt. Ranchu with Dragon Eyes has wide back and strong caudal body. The back is round and smooth of wooden comb back or long parabolic type. Later there will be enough room for the development of the goldfish with this type of back to grow a big fish. If Ranchu with Dragon Eyes' physique is not big enough, the characteristics of the first half part can not be physically and visually balanced.

红黑龙睛寿
Red Black Ranchu with Dragon Eyes

- **花色类别** / Color
  红黑 / Red Black

- **品种类别** / Species
  蛋形 > 复合变异类 > 龙睛寿型
  Egg-Fish > Compound Variation > Ranchu with Dragon Eyes

- **品种详述** / Species Description

　　图例所示为红黑龙睛寿，其外形特征完全和黑龙睛寿一样。其色为红黑，也有称之为包金色的，是黑色金鱼向红色金鱼转色的中间过渡期品种。大多数黑色金鱼都会在二龄或三龄时，向红色转变。可适当通过人为手段控制水质、温度和饵料中的微量元素延缓这个过程。红黑龙睛寿体格强健，较易于饲养。其玩赏偏向于侧视，可饲养于较深的水族箱中。

As shown in the illustration is Red Black Ranchu with Dragon Eyes. Its characteristics are totally the same with Black Ranchu with Dragon Eyes. Its color is red and black, sometimes called bag of gold, which is a variety during color transition from black goldfish turning into red goldfish. Most black goldfish are going to turn to red when it is in 2 or 3 years' stage. The color transition process can be postponed by proper artificially controlling the water quality, temperature, and the micro-element in feeding stuff. Red Black Ranchu with Dragon Eyes is strong, and is easy to rear. It is suitable for viewing from side, and can be kept in the deep water in aquarium.

花色类别 / Color •

黑白 / Black White

品种类别 / Species •

蛋形 > 复合变异类 > 龙睛寿型

Egg-Fish > Compound Variation > Ranchu with Dragon Eyes

品种详述 / Species Description •

黑白龙睛寿

Black White Ranchu
with Dragon Eyes

　　图例所示是黑白龙睛寿，是蓝龙睛寿逐渐褪色时的一个花色状态。此鱼头型丰满，两个眼睛外凸且对称，发育良好；身形短粗雄壮，身体洁白，背部有黑斑；外形有国宝大熊猫的意趣，两个黑色的龙睛正好像是大熊猫两个黑色的耳朵，十分有趣。黑白色的鱼不耐高温，高温会加速其褪色过程，因此可尽量降低饲养的水温，保持颜色的稳定。

As shown in the illustration is Black White Ranchu with Dragon Eyes which is a color of variety during the color transition period of Blue Telescope. Its head shape is plump. Two eyes are convex and symmetrical, well-developed. A short and majestic body is spotless white with black spot on the back. Its appearance is somewhat like the national treasure pandas. Its two black dragon eyes are just like the two black ears of panda, very interesting. Black and white fish is not tolerance to high temperature, which will accelerate the process of color fading, so the water temperature should be reduced as low as possible to keep the color stability.

三色龙睛寿 ——
Red Black White Ranchu
with Dragon Eyes

- **花色类别** / Color
  三色 / Red Black White

- **品种类别** / Species
  蛋形 > 复合变异类 > 龙睛寿型
  Egg-Fish > Compound Variation > Ranchu with Dragon Eyes

- **品种详述** / Species Description

　　三色龙睛寿和黑白龙睛寿一样，是蓝龙睛寿逐渐褪色时的一个花色品种。黑白色是纯蓝色转色而出的，而三色多是因蓝色鱼身上有铁锈色斑块，这些斑块转出红色，与蓝色褪出的黑白色形成黑白红三色。图例所示金鱼整体发育良好，头茸丰满、不松散，眼睛为黑色并带有虹彩，背部圆滑平顺，唯一缺憾是尾肩有些僵硬，会影响游动时的姿态。同黑白龙睛寿一样，三色龙睛寿需要降低水温来减缓褪色过程。

The same as Black White Ranchu with Dragon Eyes, Red Black White Ranchu with Dragon Eyes is a color of variety during the color transition period of Blue Ranchu with Dragon Eyes. Black and white color is a result of pure blue transition. While, the three-color is because the blue fish has rust colored patches on the body, which will turns into red, together with black and white faded from blue, are three-color: black, white, and red. As shown in the illustration, the goldfish is well-developed overall; the head pompon is plump, non-scattering; the eyes are black with rainbow and the back is round and smooth. Regrettably, the middle tail are some stiff, which will affect its swimming posture. Like as Black White Ranchu with Dragon Eyes, it is need to reduce the water temperature so as to postpone the color fading of Red Black White Ranchu with Dragon Eyes.

- **花色类别** / Color
  软鳞红白 / Red White Matt

- **品种类别** / Species
  蛋形 > 复合变异类 > 龙睛寿型
  Egg-Fish > Compound Variation > Ranchu with Dragon Eyes

- **品种详述** / Species Description

　　图例所示为樱花龙睛寿，是由樱花兰寿或五花兰寿选育而出的一个变异品种。其外部形态特征与其他花色的龙睛寿没有差异，但鳞片为软鳞，由于是幼鱼，其体色还未充分定色，红色的饱和度有些偏弱。

As shown in the illustration is Sakura Ranchu with Dragon Eyes (Red White Matt). This variation variety is a result of selective breeding Sakura Ranch (Red White Matt) or Calico Ranchu (Bluish Base). There is no difference in external characteristics between Sakura Ranchu with Dragon Eyes (Red White Matt) and other color variety of Ranchu with Dragon Eyes. But its scales are soft. Because it is the baby fish, its body color has not been set yet and the color saturation is a little bit weak.

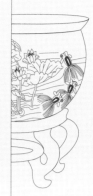

蛋形
Egg-Fish

∨

复合变异类
Compound Variation

∨

望天球型
Celestial-Eye with Pompons

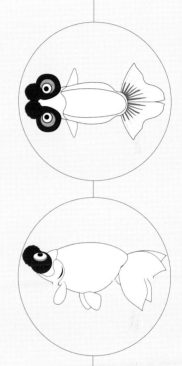

　　望天球，是望天中带绒球的一个复合变异类品种。望天球在民国时期的资料中即有记述，是中国金鱼中较早的一个品种类型。望天球属于小众品种，很少有鱼场饲养，目前主要产地为福州，苏沪皖地区也有部分鱼场养殖。望天球中红色最多，其他花色还有黑望天球、紫望天球和红白望天球。近几年，北京的金鱼爱好者又重新培育出了蓝望天球、雪青望天球、五花望天球等新的花色品种。望天球体形要求短圆粗壮，圆鳍小尾，尾芯与尾筒接合处不能有尾刺；眼外凸上翻到位，且要对称，绒球紧实不松垮。望天球和望天一样，适合俯视玩赏，以大木海或陶缸饲养一群，十分赏心悦目。操作望天球时要小心，绒球一旦损伤，虽可自愈，但要想和原来一样，需要花很长的时间。绒球需要适当修剪，防止其变松或走形。

Celestial-Eye with Pompons is a compound variety of Celestial-Eye with Pompons. Celestial-Eye with Pompons is a small-popularized variety and it is reared by few fish farms. Recently, it is mainly reared in Fuzhou, or some fish farms in Suzhou, Shanghai and Anhui area. Most of Celestial-Eye with Pompons are red and there are also some other colors: Black Celestial-Eye with Pompons, Chocolate Celestial-Eye with Pompons and Red White Celestial-Eye with Pompons. In recent years, some new breeds such as Blue Celestial-Eye with Pompons, Lilac Celestial-Eye with Pompons, Calico Celestial-Eye with Pompons (Bluish Base) has been cultivated by goldfish enthusiast in Beijing. Celestial-Eye with Pompons has short, round and thick body; round fins and small tail. There should be no caudal spine on the junction of tail center and caudal body. The eyes are convex and upturning to the position and symmetric. The Pompons is tight, non-scattering. Like Celestial-Eye, Celestial-Eye with Pompons is suitable to overlooking view. Breeding a group of Celestial-Eye with Pompons in "Da Mu Hai" (a kind of wooden cylinder) or ceramics cylinder, it is pretty charming. It should be very careful when in operation. Once harmed, although the Pompons can be self-healing, it will take a long time to bounce back to the original. The Pompons need to be properly trimmed to prevent it from loosing or scattering.

- **花色类别** / Color

  红 / Red

- **品种类别** / Species

  蛋形 > 复合变异类 > 望天球型

  Egg-Fish > Compound Variation > Celestial-Eye with Pompons

- **品种详述** / Species Description

  图例所示为当岁的红望天球，全身火红，背幅宽，尾柄粗壮有力，圆鳍小尾；眼睛外翻对称，与头顶平齐，绒球紧实对称；背部平顺度略有欠缺。红望天球体质强健，比较易于饲养。其生性活泼好动，与人的亲和力互动性均较好。

  As shown in the illustration is Red Celestial-Eye with Pompons in its first year stage, totally red as fire. Its back is wide, caudal body is thick and strong, with round fins and small tails. The eyes are ectrophy and are symmetrical, flush with the top of the head. The Pompons are tight and symmetrical. The back is not so smooth. Red Celestial-Eye is strong physically and easy to rear. It is lively and active, full of affinity and interaction with people.

白望天球

White Celestial-Eye with Pompons

- **花色类别** / Color
  白 / White

- **品种类别** / Species
  蛋形 > 复合变异类 > 望天球型
  Egg-Fish > Compound Variation > Celestial-Eye with Pompons

- **品种详述** / Species Description

　　图例所示为白望天球，是红白望天球中出现的全白个体。一般全白的个体行话叫做"白皮"，很少留下饲养，但是在培育红白花的金鱼花色时，适当使用白色鱼做亲本，可以提高红白色的比例，并且会出现好看的花纹。白望天球进入二龄期以后，头部及绒球会变为淡淡的柠檬黄色，颇显素雅别致，如果配上眼睛带有虹彩的三环套月或朱砂眼，就更为难得了。

As shown in the illustration is White Celestial-Eye with Pompons which is an individual totally white from Red White Celestial-Eye with Pompons. Generally, a total white individual is called "white skin" which is seldom left to rear. It can be used as parents when people cultivating the color variety of Red White Calico Goldfish to increase the proportion of red and white for obtaining beautiful color pattern. After White Celestial-Eye with Pompons grows into the second year stage, its head and Pompons will turn into a touch of lemon yellow, very elegant chic, and if accompanied by the eyes with a rainbow of moon in three rings or cinnabar eyes, it is even more rare.

花色类别／Color •

黑／Black

品种类别／Species •

蛋形 ＞ 复合变异类 ＞ 望天球型

Egg-Fish ＞ Compound Variation ＞ Celestial-Eye with Pompons

品种详述／Species Description •

黑望天球

—— Black Celestial-Eye with Pompons

　　黑望天球是红望天球里未转色的个体，有的个体甚至会保持二三年不变色，黑望天球在外部形态上与其他花色的望天球完全一样。黑望天球的转色过程也十分有看点，有的个体会褪色为两个红球的朱球黑望天，有的会褪色为整个头部是红色而全身墨黑的红头墨身望天球，虽然不能稳定，但也给玩赏者带去惊喜。

Black Celestial-Eye with Pompons is an individual without color transition in Red Celestial-Eye with Pompons. Some individuals will keep this color for 2 or 3 years. Its characteristics are the same as that of other color varieties of Celestial-Eye with Pompons. It is very value to watch the process of the color transition of Black Celestial-Eye with Pompons. Some individuals will fade into two red Pompons called Black Celestial-Eye with Red Pompons, and some will fade into Celestial-Eye with Pompons with totally red head and black body. Although it is unstable, it can bring surprise to the viewers.

蓝望天球
—— Blue Celestial-Eye with Pompons

- **花色类别** / Color
  蓝 / Blue

- **品种类别** / Species
  蛋形 ＞ 复合变异类 ＞ 望天球型
  Egg-Fish ＞ Compound Variation ＞ Celestial-Eye with Pompons

- **品种详述** / Species Description

  蓝望天球是较古老的一个品种,笔者2005年曾在杭州鱼展上见过两尾,而后这个品种就慢慢消失在人们的视线中。近两年来,北京的一位金鱼爱好者通过杂交手段,重新选育出了蓝望天球。现在的蓝望天球身形比以前的要偏长一些,显得有些文弱,也有可能是养殖手法造成的差异。蓝望天球中还会褪色为黑白色或喜鹊花的个体,目前还在稳定培育中。

  Blue Celestial-Eye with Pompons is an ancient breed. Since the author watched two of them in Hangzhou Fish Fair in 2005, this species have slowly disappeared in the line of people's sight. In recent two years, Blue Celestial-Eye with Pompons has been re-cultivated through hybrid breeding by a goldfish enthusiast in Beijing. The Blue Celestial-Eye with Pompons at present is a little bit longer than that of Blue Celestial-Eye with Pompons in the past. Blue Celestial-Eye with Pompons at present seems somewhat frail maybe it is because of the different breeding practices. The black and white colored or magpie flowers individuals can also be faded out from Blue Celestial-Eye with Pompons, but now they are cultivated stably.

花色类别 / Color •
紫 / Chocolate

品种类别 / Species •
蛋形 > 复合变异类 > 望天球型
Egg-Fish > Compound Variation > Celestial-Eye with Pompons

品种详述 / Species Description •

紫望天球 — Chocolate Celestial-Eye with Pompons

　　紫望天球也是望天球中出现较早的花色鱼种，是福州金鱼业者从红望天球中选育出的。紫望天球身形粗壮，眼睛经常会出现红色的虹彩，其球紧实，有时会出现紫身双红球的个体，非常珍贵。由于它是从红望天球中选育出的，因此会出现转色成红望天球的个体，且随温度上升，其变色比例愈高。但随着金鱼业者不断提纯选育，其颜色越来越稳定，是福州金鱼又一个优秀代表品种。

Chocolate Celestial-Eye with Pompons is also an earlier variety in Celestial-Eye with Pompons family, which is the result of selective breeding Red Celestial-Eye with Pompons, by Fuzhou goldfish industry. Chocolate Celestial-Eye with Pompons 's stature is stout and there is always red rainbow in its eyes. Its Pompons are tight and sometimes there is an individual with Chocolate body and double red Pompons, very rare. Because it is cultivated from Red Celestial-Eye with Pompons, it can turn into the individual of Red Celestial-Eye with Pompons. And with the increase of temperature, the proportion of color change is growing higher. With continuous purification and selective cultivation, its color becomes more and more stable. It is another excellent representative variety of Fuzhou goldfish.

花色类别／Color •

雪青／Lilac

品种类别／Species •

蛋形 ＞ 复合变异类 ＞ 望天球型
Egg-Fish ＞ Compound Variation ＞ Celestial-Eye with Pompons

品种详述／Species Description •

　　图例所示为雪青望天球的亚成体，雪青望天球属于小众花色品种，由于培育时间不长，其正品率较低，因此非常稀有。雪青花色可以从蓝色、紫色金鱼中选育。图例金鱼全身呈淡雪青色，鱼鳍边缘点缀有少许紫色斑块，尾鳍是连尾，在中国金鱼的品评标准中属于硬伤。雪青望天球的球体发育较迟，二龄以后才逐渐丰满。但雪青色较为稳定，褪色的比例较少。

As shown in the illustration is sub-adult Lilac Celestial-Eye with Pompons, which is small popularized variety. Because the cultivation time is not long, the genuine rate is low, and it is very rare. Lilac color can be selective cultivated from blue and Chocolate goldfish. The gold fish shown in the illustration is totally pale Lilac color, with few Chocolate patches decorated on the fins' edge. Its caudal fin is connected with tail, which is flawed in Chinese goldfish criteria.

雪青望天球——Lilac Celestial-Eye with Pompons

红白望天球
— Red White Celestial-Eye with Pompons

- **花色类别** / Color

  红白 / Red White

- **品种类别** / Species

  蛋形 > 复合变异类 > 望天球型

  Egg-Fish > Compound Variation > Celestial-Eye with Pompons

- **品种详述** / Species Description

　　红白望天球是望天球家族中最受大众欢迎的一个花色类型。红白望天球体形和红望天球完全相同。其花色变化十分丰富，有船底红、玉顶、玉面、玉带、蟒纹、鹿子等等。红白望天球以双红球为贵，鸳鸯球次之，双白球则再下一等，眼睛要上翻至头顶，且大小对称，如果能出带虹彩的三环套月则更为难得。红白望天球体质强健，抗病力也较强，比较容易饲养。近些年来，红白望天球在金鱼爱好者中逐渐流行开来。

Red White Celestial-Eye with Pompons is the most popular color of variety in the Celestial-Eye with Pompons family. It is rich in color change, there are belly red, jade head, jade cheek, jade belt, python pattern, scarlet scale and so on. In Red White Celestial-Eye with Pompons, the double red Pompons are the best, followed by different colored Pompons and the last is double white Pompons. The eyes should turn over to the top of the head, and they are should be symmetric in size, and if with a rainbow of moon in three rings, it will be even more rare. Red White Celestial-Eye with Pompons' physique is strong and be able to resistant to the disease so it is easy to breed. In recent years Red White Celestial-Eye with Pompons has been gradually popular with goldfish enthusiasts.

朱砂望天球
Vermilion Celestial-Eye with Pompons

- **花色类别** / Color

  红白 / Vermilion

- **品种类别** / Species

  蛋形 > 复合变异类 > 望天球型

  Egg-Fish > Compound Variation > Celestial-Eye with Pompons

- **品种详述** / Species Description

  图例所示为朱砂望天球。朱砂望天球是红白望天球中的极品花色，与龙睛蝶尾中的十二红、水泡中的朱砂水泡一样，都不具色彩遗传性，是一种巧色。朱砂望天球全身洁白，没有一点杂色，唯独两个绒球呈鲜红色。鱼嘴上方有一小抹红色，如同擦了口红，十分俏皮。同样的红白巧色望天球花色还有：双红眼的朱砂望天球、红头红尾的首尾红朱砂望天球，都是可遇不可求的花色品种。

As shown in the illustration is Vermilion Celestial-Eye with Pompons, which is the highest grade color in Red White Celestial-Eye with Pompons. Like twelve red in Butterfly Moor and cinnabar Bubble-Eye in Bubble-Eye, the color is formed by chance without color heredity. Vermilion Celestial-Eye with Pompons is totally spotless white, without any variegate patches except for two bright red pompon Pompons. There is a small wipe red on its mouth, as if wiped with a lipstick, very cute. There are other colors of Celestial-Eye with Pompons, of which the color is also formed by chance, such as Vermilion Pompons with double red eyes and Vermilion Celestial-Eye with Pompons with red head and tail. All of them are lucky color varieties by chance.

红黑望天球
—— Red Black Celestial-Eye with Pompons

- **花色类别** / Color
  红黑 / Red Black

- **品种类别** / Species
  蛋形 ＞ 复合变异类 ＞ 望天球型
  Egg-Fish ＞ Compound Variation ＞ Celestial-Eye with Pompons

- **品种详述** / Species Description

  图例所示为红黑望天球，是黑望天球向红望天球转色的一个过渡期品种。由于鱼自身的内在差异和外界环境差异，这个过程长短不定，笔者曾经饲养过一尾黑望天球，是4年都没变色的个体。红黑色虽是一个过渡阶段色，却让饲养者了解到金鱼不断变化的魅力。红黑望天球体质强健，适合俯视，比较易于饲养。

  As shown in the illustration is Red Black Celestial-Eye with Pompons, which is a variety during the color transition from Black Celestial-Eye with Pompons to Red Celestial-Eye with Pompons. Due to the inherent differences in the fish itself and the external environment, this process is uncertain. The author has reared a Black Celestial-Eye with Pompons, which is a individual that has not changed its color for 4 years. Although the red and black color is a color of transitional stage, the breeder can understand the changing charm of goldfish. Red Black Celestial-Eye with Pompons is strong, and suitable for overlooking and easy to rear.

五花望天球 — Calico Celestial-Eye with Pompons (Bluish Base)

- **花色类别** / Color

软鳞五花 / Calico (Bluish Base)

- **品种类别** / Species

蛋形 > 复合变异类 > 望天球型
Egg-Fish > Compound Variation > Celestial-Eye with Pompons

- **品种详述** / Species Description

　　图例所示为五花望天球亚成体，全身素兰花，带有黑色斑点，眼睛上翻对称，且有近似三环套月的虹彩。五花望天球早年曾是福州金鱼的名贵品种，后来一度灭绝，目前的五花望天球是金鱼业者近几年来重新杂交培育的，其饲养数量不多，较为珍稀，目前只有北京、山东、安徽和江苏部分鱼场有少量培育。五花望天球体形强健，在望天球中可以培育出较大的个体，但其球体发育较迟，要二龄后才能达到标准。

As shown in the illustration is Vermilion Celestial-Eye with Pompons, which is the sub-adult Calico Celestial-Eye with Pompons (Bluish Base). It is totally vegetarian orchid, with black spot. Its eyes turn over and is symmetric with a rainbow of moon in three rings. In early years, Calico Celestial-Eye with Pompons (Bluish Base) has ever been rare varieties of Fuzhou goldfish, and has once been extinct. Calico Celestial-Eye with Pompons (Bluish Base) at present is re-cultivated by goldfish industry. The breeding quantity is not so much, so it is very rare. At present it is reared in only few fish farm grounds in Beijing, Shandong, Anhui and Jiangsu. Calico Celestial-Eye with Pompons (Bluish Base)'s physique is strong, which can grow in a big size while is Pompons developed late. After 2 years stage it can grow to the standard.

中国金鱼鉴赏

文 / 汪聿钢 陈镇平

　　尾展拖轻縠，妆新炫绛霓，中国金鱼迄今已有千余年历史，自古以来深受各国人士喜爱。中国金鱼被美誉为水中牡丹，美名遍及世界，因其体形丰满、性情温和，也被誉为幸福、吉祥、友谊与和平的象征；其所表现出的绰约风姿、绚丽色彩、祥和安逸给人带来的更是一种质朴、娴静而优雅的意境，是活的艺术品。

　　在人类文明史上，中国金鱼承载着深厚的华夏文化，被誉为"东方圣鱼""华夏瑰宝"，是中国的国粹之一。国鱼形象影响着中华文化的传播，目前世界各地的金鱼品种都是直接或间接由中国引入的，也是世界观赏鱼史上最早的人工选育品种。其品种多样、形态各异、泳姿舒逸，更有着无穷无尽的变化，也让许多人悠然神往、陶醉其中。

　　仪态万千的金鱼呈现之美不尽相同，审美标准则仁者见仁，智者见智。审美因涉及对金鱼的认识程度、审美意识及个人喜好等因素而有所不同，众说纷纭之际，也尚未有达成业界共识的判定标准。概括来讲，审美标准的差异性受几点因素影响：

　　时代空间之差所带来的审美之别，在于审美活动因时代不同可产生不同之美的结果。审美感受既受评审者"自我价值取向"的影响，又受其所处的时代、民族、文化等环境的制约，而产生时代性的差异美感，相同特点也存在于服装造型等人类文化艺术的审美上。

　　民族文化之异所带来的审美之别，在于同一民族受着相同环境条件的影响，必然在审美行为中表现出某种主流的偏好。就此因素相对其他民族而言，便构成独特鲜明的民族特色，亦导致中国金鱼因得益于包容性的文化而纷呈数百品种。

　　市场因素所带来的审美之别，在于金鱼品系形态各异，在市场上受追捧程度不一，受市场供求关系拉动始终呈动态变化的状态，迎合市场的审美喜好便成了对金鱼品种取舍与否显著特征之一。这是金鱼本身所具备的美学特质和不同评审者的审美判断所产生的差异结果。

个人主观偏好不同所带来的审美之别，在于审美之中伴随个人情感色彩。面对同一尾金鱼，每个人所产生的审美感受往往不完全一样。作为不断变化着的人和社会客观环境，人的审美观念也是不断变化着的。这也是审美行为中回避不了的个体差异性，即是美感的主观性。审美行为的这种个性差异皆因评审者的审美能力、文化艺术修养、经验、思想感情等等多个因素影响所形成。

即便金鱼审美有着多种差异，在我们的审美活动中又存在着时代的、民族的、地域的、个体的差异，但我们不应且难以将这种差异绝对化。事实上即使在不同时代、不同民族、不同个体的审美者，在一定时间范围内对同一审美对象往往仍能找到一些相近或相似的审美感受，这便是审美意识中的共同性。

浅谈金鱼共性乃需观其"好之美""大之美"及"稀有之美"，三者的协调统一为高质金鱼审美标准。只"好"不"大"、不"稀有"，这种美有不足；光"大""稀有"不"好"，这种美也有瑕疵。

"好之美"则表现在金鱼形态、色彩与运动神韵等方面。形态之美皆为身形健全端正，尾柄粗壮，尾鳍对称、夹角合理、自然张开、既不低垂也不上翘。此类金鱼无论是游动或是静止时，都能保持鱼体的平衡和稳定。若尾柄瘦小，尾鳍下垂，不仅尾鳍摆动费力，而且尾鳍摆动时所产生的推动力与鱼体中轴线不在同一水平线上，使金鱼的游姿呈俯冲状，甚至会尾朝上，头朝下，形成"倒栽葱"。反之，如果尾鳍上翘，尾鳍摆动时所产生的推动力使金鱼呈向上冲刺状，所有这些都会影响金鱼游姿之美。

各鳍发达、自然舒展，没有明显的卷折。有背鳍者背鳍宜挺拔，无背鳍者背部曲线应流畅优美，背上不能有残鳍遗痕。其鳞片沿鳃盖后缘至尾鳍基部，从背脊到腹底，排列均匀紧凑、整齐划一即为上品。反之，如有鳞片脱落或大小不一、排列参差不齐，当属次品。

　　其他部位均衡、对称，鱼体健壮、匀称，其吻、眼、鼻、鳃、鳍、躯干等身体各部位比例均匀，游姿进退自如。如望天类的眼睛朝天方向左右应一致；水泡类的水泡、绒球类的绒球应紧致结实而对称；高头类的肉茸方正厚实；文鱼体短背高，头尖如鼠，整个身体的形状如两个拼合的三角形；龙睛身体各部位包括鳍基在内几乎都是一个个大大小小的圆柱体，各个部位大小、长短比例相称，给人丰满圆润的感觉；负重型的金鱼如水泡、虎头等，背部顺滑、腹部浑圆、身体粗壮，因其粗壮，身体平衡能力则更强，其头部的负载量就更大，其品种特征就更强烈突出。这一赏鱼之道在我国古代已有印证，如清朝末年姚元之《竹叶亭杂记》中记载的"身粗而匀、尾大而正、眼齐而称、体正而圆"。

　　金鱼的品种特征强烈、突出，对于不同品种特征的鱼，还能满足爱好者不同的需求。如高头类金鱼，背鳍、头肉高大，尾鳍宽长，游姿稳健，很有气魄；兰寿、寿星类则如金鱼王国中的"力士"，呈现出强劲的刚阳之美；珍珠鳞体形浑圆，鳞片如珍珠般饱满，粒粒可数；望天眼眼如铜铃，双目直线朝天。而对于两个品系金鱼杂交而产生变异的金鱼，更应有强烈突出的品系特征，其品系特征越强烈、越突出就越好。但若特征过分突出，而引致身体部位结构比例失调，造成身体重心前后倾斜，则显得不伦不类而影响其形态之美。如水泡，当其身体难以承受头部重负之时，常静伏于鱼缸底部，因而影响了它的观赏价值。

　　基于金鱼色彩而言则分为单色复色二类。单色则青、白、红、黑、蓝、紫，复色则有红白、红黑、黑白、紫蓝、三色、五花、虎纹等。其白、青皆为平淡，若非特征强烈突出，则为淘汰对象。单色金鱼鲜艳夺目，复色金鱼对比鲜明。如黑狮头黑色浓烈方为上品，黑色呈青色且混浊则缺乏魅力；红白、红黑、黑白、紫蓝等复色，体色对比鲜明且相衬尤为吸引人。

相对来说，标准的审美以金鱼的形体、色彩对称为佳品。但是有些金鱼的眼睛或是金鱼的水泡左右色彩不一，因此被称之为鸳鸯眼、鸳鸯水泡，此类金鱼若非色彩配搭十分巧妙而不可多得的，一般来讲观赏价值不高。

值得一提的是，随着鱼龄增长鱼体部分器官色彩比例的变化各有不同。原本个体特征不突出者可能因鱼龄增长而有所改善，其黑白、紫白、蓝白、紫红、红黑、三色等金鱼品种也非永久稳定，色彩或因水质、光照、营养以及温度变化而改变甚至消失。如俗称铁包金的红黑类花色，在幼鱼时期身体呈黑色，随着鱼龄的增长及条件的转变，其黑色会从腹部开始逐渐退却，出现红色，最终变成全身红色。

一般而言，鱼龄于一岁之内的金鱼正处于生长发育高峰期，鱼体比例也处于快速变化之中，因此这一时期的金鱼美之特征存在不确定性，仍可能发生出乎意料的变化。而鱼龄于一岁以上者，鱼体生长发育速度明显减慢，形态特征已相对稳定。因此一冬龄以上的金鱼所具有美的特征比当岁鱼龄的金鱼更为持久可靠。

游动时的金鱼柔美有力，姿态洒脱或是轻纱飞舞，纤丽飘逸；静止时沉浮轻巧，一个细小动作，皆蕴含意境之美。金鱼有了运动之美，犹如舞蹈乐动的艺术，才会产生无限情趣韵致，而这种情趣和韵致也便构成了金鱼的神韵美。

"大之美"则体现在金鱼整体形态方面，皆为同品系、鱼龄及体长体重作比较。以金鱼群体内部品系而言，有些品系的体格相对较大，如高头类、琉金类等；而有些品系的体格相对较小，如水泡、望天、蝶尾、绒球等。品系各有大小，皆由各自特性所决定，若取不同品系的金鱼作比较便会混淆评价之标准。

正常生长的大体形金鱼并正值年轻之时才堪称大之美。若已自然长大但已临近暮年，生长潜力几近枯竭则是一种缺陷之"大"。与老龄鱼相比，必然越年轻评价越高。

一般来说，对于鱼体体长及体重衡量则分为三大准则：一则身体全长相同且体长体重相同之下，其审美考量相同；二则身体全长相同但体长体重不同之下，以其体重大者为高审美考量；三则身体全长不同但体长体重相同之下，其审美考量相同。

"稀有之美"意即物以稀为贵，因而"稀有"亦为金鱼鉴赏评价的重要因素。如现有金鱼品系中，形体、色彩、运动及神韵皆备之时，既"好"且"大"的则少之又少。为此，具备上述因素同时也具备"稀有"之美的金鱼倍显珍贵，因珍稀独特而意味市场价值更高。

另一方面，大自然的发展有它必然的规律，人类在遗传学的所有成就最终也无法完全脱离生物与自然界的自主性。我们能够定向改变有机体的品性，培育出合乎当前审美或引领潮流的新品种，但却永远不能够完全控制生物的发育及发展。金鱼形体的残缺是每个金鱼养殖者不希望看到的，却又是不能避免的自然作用的结果，由此徒生许多遗憾。在定向积累有利变异品种的过程中，偶然会遇见因突变产生的奇特形态个体，这种突变个体的出现未必尽是无价值的品种，甚至因其品种特征强烈突出还可满足不同爱好者之需求。

金鱼养殖生产力的发展水平也会影响到审美的客观标准，它与审美标准之间，是一种互相依存、互相促进、共同发展的关系。每一个养殖者在"创造"新品种的过程中，都必然要受到现有客观审美标准的影响。随着科技日新月异，金鱼饲养技术日渐提高。中国金鱼在人工饲养的条件下历经千余年的发展、演变，形成了具有浓厚自身特色、变异最多的观赏鱼体系，深得世界各届人士喜爱，在中国金鱼产业飞速发展的今天，更为世人所看重。历代的金鱼养殖家们为了实现当时人们审美爱好的需求，对金鱼不断进行倾向性的选育，淘汰不利于变异的个体，变异愈严重，就愈为养殖者所喜爱而予以保留，使得现有的金鱼变异很大。

纵观金鱼的演变历史，通过人工定向选择及变异，人为地发展了500多个金鱼品种(据图鉴最终专家共识)，"美"作为一种人类意识，它也是不断随着时代的变换、品种的演化而发生着变化。这完全受人及种种因素的作用，经过漫长岁月的变异积累，便有了今天千姿百态的金鱼品系。也因此，被定义"稀有"特质的金鱼并非一成不变，但也并非通过大量繁殖或技术改造就可获得。一旦某种"稀有"金鱼得到大量繁殖便失去其"稀有"之相应价值。

综上所述，金鱼审美因应时代特征、民族特色、个性特点而不尽相同，却又在一定时间和地域范围内形成趋同的审美趣味、偏好与理想。审美意识作为人类历史发展的一种积累和沉淀，它既有历史继承性的一面，又饱含着时代创新的趋势。所以，金鱼审美的客观标准虽是客观存在着，但这种客观标准绝非一成不变，它随着客观环境的变化而演化和发展，直接受到人类生活发展演变的影响，不断有新的、更符合时尚品味的品种出现，相应的审美判断标准也就在逐渐变化。

金鱼进化永不停息，美态也因此无穷无尽，可以说发现与创新是成为这种文化源远不息的永恒主题。金鱼鉴赏，并无最好，只有更好！

Appreciation of Chinese Goldfish

Article / Wang Yugang　Chen Zhenping

The tail is swaying like the elegant willow; while its colorful "dress" is extraordinarily dazzling. Chinese goldfish owns a history of over thousands of years, and is well received in many countries. It has gained a worldwide fame-the "peony in water" out of its plump body and moderate temper. It is also regarded as a symbol of happiness, good luck, friendship and peace. The graceful gestures, colorful bodies and soothing image of the goldfish bring to us a kind of simple, quiet and elegant mood. It is so called the "Living Art".

In the history of human civilization, the Chinese goldfish carries a profound cultural significance. As image of the national fish, it helps promote the spread of Chinese culture to all over the world. At present, the worldwide goldfish varieties are directly or indirectly introduced from China and have accompanied the people of all nationalities living thousands of years. Chinese goldfish are the earliest artificial cultivation varieties in the ornamental fish history. Reputed as the "Oriental holy fish", "Chinese treasures", they are listed as the quintessence of China. With varieties of species, diverse shapes and changing gestures, goldfish are so attractive and enchanting that many people are intoxicated into it.

Despite of multifarious shapes, the beauty of goldfish is identically recognized. But different people have different views. Specifically, it depends on individual's cognitions of goldfish, aesthetic sense, personal preference, so on and so forth. Opinions vary, so the consensus of industry standard of aesthetics has not been reached yet. Generally speaking, the difference of aesthetic standard is influenced by the following factors:

The time-difference aesthetics can result in different aesthetic perceptions, which are affected by reviewers' self-value orientation, the age they are living in, nationality, culture and other constraints factors. Likewise, this is also true on difference of human's culture, art, and clothing caused by time.

Secondly is the culture-difference aesthetics. Living in the same nation with same environment condition, people's aesthetic is bound to show the same preference for a certain mainstream. Compared to other nations, the unique and distinctive national features are formed. That is why there are hundreds of varieties of Chinese goldfish of different nations.

Thirdly is the market-difference aesthetics. There are numerous varieties of goldfish, their popularity are differentiated in the market, and keeps changing due to the change of supply-and-demand needs. Therefore, catering for market becomes one of the remarkable features on the variety selecting of the goldfish. This is resulted from different aesthetic property of goldfish itself and variant aesthetic judgements of the reviewers.

The fourth is the preference-difference aesthetics. This means different individual has different feelings towards one specific goldfish due to personal emotional factors. People's aesthetic also keeps changing with their mindset and social environment. It reflects the individual variation and subjectivity of aesthetics that cannot be avoided during appreciation activities, which is influenced by artistic ability, art culture, personal experience and emotions, to name just a few.

Given the diversified aesthetics differences on goldfish, and differences of appreciation activities due to factors from eras, nations, regions, and individuals, we shall not and cannot absolutize them. In fact, reviewers from different eras, nations sometimes have identical appreciation feels, which is regarded as the commonality in aesthetic cognition.

Commonly, "beauty of verve", "beauty of large", and "beauty of rare variety" are the three main principles for the industry standard of best goldfish appreciation, they are complementary to each other. Lack any of them, the goldfish cannot be considered perfect.

"Beauty of verve" refers to perfect combination of shape, color and movement, such aspects and so on. A nice goldfish must have a sound body with strong caudal body and symmetrical caudal fin, opening naturally in a reasonable angle, neither drooping nor cocking upward. This type of goldfish can keep body balanced and steady when swimming and resting. On the contrary, goldfish with small caudal body and caudal fin that are bent down, has difficulty to keep their body axis at the same level as the impetus which is created during the caudal fin is swaying, due to this, the goldfish always keeps diving gesture, with tail up and head down, just like "falling head over heels". The other way round, if the caudal fin cocks, it boosts up the body too much. All in all, these elements really can impact goldfish swimming gesture.

Goldfish is rated as delicate if it grows with well-developed fins, spreading naturally, without obvious convolution. Fins must be tall and straight. If there's no dorsal fin on a goldfish, the back has to be streamlined, broken fin residue are is not allowed. For the scales, they must be tidy, compact and uniformed, extending from operculum to the end of the caudal fin, reaching from back to the bottom of its belly. Vice versa, the goldfish can be degraded.

Its snout, eyes, nose, gills, fins and body should be well proportioned. It helps goldfish swim freely. For example, the eyes of Celestial-Eye are symmetrically positioned; the pompon of Bubble-eye, Pompon Type, are compact and balanced, the flesh grows of Tall-Head is square and thick. With short body, Fantail Goldfish has high back and a mouse-like head. Its shape is just like two triangle combining together. Every part of Dragon Eyes, including fin root, is as big or small cylinders in perfect size and height, leaving people a feeling of fullness and roundness. Heavy-duty goldfishes such as Bubble-Eye and Tigerhead have slide back, round abdomen, and strong body. Attribute to these features, the balance capability is better for them and the load capacity of the head becomes much stronger, which makes their varietal characteristics more outstanding. In fact, this appreciation way has been applied since ancient times, like the expressions "strong and even body, big and upright tail, uniform and symmetrical eyes, straight and round form", expressed in *Bamboo Pavilion Stories* by a person called Yao Yuanzhi in late Qing Dynasty.

With unique and outstanding characteristics, different varieties can meet different needs of enthusiasts. For example, Tall-Head goldfish, with dorsal fin, high sarcoma on the head, wide and long caudal fin, steady swimming posture, looks really very charming; Ranchu, Lionhead (Southern), like "herculean" in the kingdom of goldfish, show people with the beauty of strength with gorgeous plump scales; Celestial Eye has bell-like upturned eyes. Goldfish that are hybridized from two varieties are usually more characteristically distinctive, just to be moderate, not too much, otherwise, it may lead to the body gravity tilts, which will impact the beauty of appearance. Take Bubble-Eye as an example, if the head is oversized than the body can bear, they will be forced to stay on the bottom of fish tank, thus affecting the ornamental value.

According to color category, the goldfish can be divided into two classes, monochromatic color and compound color. The first one refers to cyan, white, red, black, blue, and purple, while the second refers to red white, red black, black white, purple blue, tri-color, calico, tiger streaky and so on. Among these colors, cyan and white are common colors, so very easy to be ignored if having no special characteristics. Monochrome is attractive bright, while compound colors have sharp contrast. For example, Black Lionhead is regarded as top level if it is jet-black; otherwise, it won't be so appealing. Comparatively speaking, the compound colors such as red-and-white, red-and-black, black-and-white, purple-and-blue and so on, are normally more attractive to people.

In normal sense, per industrial aesthetics standard, top grade goldfish are those with symmetrical shape and perfect color mixture. However, sometimes some goldfish grows with asymmetry eyes or colors of bubbles. This variety will be cherished only if the color is perfectly matched. People give them beautiful names as "Yuan yang" eyes or "Yuan yang" bubbles. (Yuan yang, Mandarin Ducks, used as objective to describe asymmetry object.)

It is interesting to notice that with age growing, the body color changes. Therefore, the goldfish with no special characteristics may be improved a lot when growing up, meanwhile, varieties such as black white, chocolate white, blue white, chocolate red, red black, tri-color and so on, are permanent stable, they may change into another color or fade away due to such elements like water quality, illumination,

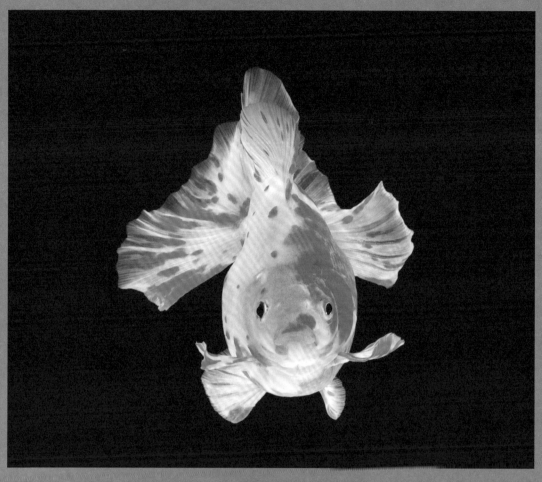

nutrients and temperature and so on. For example, the red-black kind has black body in its young age. With the change of age and conditions, the black is gradually faded from abdomen while the red is developed and extend all over the whole body.

Generally, the goldfish grows fast in their first year, the body size can develop rapidly. During this period, any kind of unexpected changes may happen. But one year later, the growing speed of goldfish will obviously slow down, the steady morphological characteristics will be showing up. So the goldfish beyond one year old has more stable and reliable beautiful characteristics than the one within one year old.

When swimming, the goldfish is soft but vigorous, as graceful as the dancing lady; when resting, it is light and handy, so charming even with a tiny movement. The beauty of movements is like the art of dancing, thus bringing people with endless interests and enchantments, just reflecting the verve of the goldfish.

"Beauty of large" refers to morphology of goldfish, differentiated among varieties, ages, lengths and weight. It has to be judged within the same variety. On different varieties, some are large sized, such as Tall-Head and Ryukin; some are small, such as Bubble-Eye, Celestial-Eye, Butterfly Moor, Pompon and so on; Their size are determined by their variety characteristics. So we cannot use one same standard to judge different varieties of goldfish.

Goldfish can be called "beauty of large" only when it is young. Large but old, merely reaching the end of the life, this kind of goldfish has no potential anymore. Therefore it is called large with defective. The younger the goldfish is, the higher rating it will possess.

Generally speaking, there are three standards to measure the length and weight of goldfish: first, the judgment of aesthetics on goldfishes with same length and same weight are same; second, for goldfish with same length and different weight, the judgment of aesthetics on the one with heavy weight is much higher; third, the judgment of aesthetics on the goldfishes with different fall length but with same body length and weight are also same.

"Beauty of rare variety" means "the thing that is rare is dear". So "rarity" is the one of the key factors on appreciation of goldfish. Among current goldfish varieties, it is not so common to see charming, large-sized goldfish with all the best features like perfect shape, gorgeous color, and great vitality. Therefore, commercially, when attached with a tab of "rarity", it will become extremely valuable.

On the other hand, everything has its laws. All human's achievements in genetics cannot be separated from the autonomy of creatures and the nature world. We may make oriented change on the characters of organic entity, but never can we fully control the growth and development of creatures. The deformity of goldfish, although is unexpected by cultivators, is an unavoidable per law of nature. Being miserable is making no sense, because man can cultivate new variety but cannot never fully manipulate the growth and development of it. During the process of accumulating favorable variational variety, some entity with strange morphology may appear. The variation is not completely valueless, sometimes, it may meet different fanciers' needs due to its distinctive features.

The cultivation development of goldfish is also affecting aesthetics standards, they are inseparable, mutual promoting to each other. During the process of "creating" new variety, the cultivator is always influenced by current aesthetic standards. With the rapid development of technology, the technique of cultivating goldfish increases gradually. After thousands of years' artificial breeding and numerous, Chinese goldfish has formed an ornamental species with unique characteristics and various vibrations, which is favored by people from different industries in the world, especially for today when goldfish industry is blooming so fast in China. In order to meet people's requirements on aesthetics, generations of cultivators breed goldfish with orientation to eliminate individuals against vibration and reserve ones with highly vibration, thus resulting in huge vibrations on current goldfish.

Throughout the history of goldfish evolvement, more than 500 hundreds varieties have been developed by artificial oriented selection and natural variation (recognized by experts of this book). The definition of "beauty", as one of human's consciousness, is constantly changing with eras and among different varieties. After many years of accumulated variation and influenced by man and other factors, thousands of goldfish varieties come into being. So the goldfish defined as "rare" is not unalterable, and cannot been produced by mass propagation or technique transformation. Once the goldfish with "rare" characters gets mass propagation, its value of "rarity" will be lost.

In conclusion, the aesthetics of goldfish are not always the same because of the differences on eras, nations and characteristics, while convergent aesthetics, preference and ideal may be formed in specific time and region. As an accumulation and sediment in the history of human being, aesthetics is inherited in history and embodies the trend of innovation. So the objective standards for aesthetic of goldfish may exist, but they are changing with the development of environment. Affected by the changes of human being's life, new varieties keep appearing to fit the trend and fashion. So the aesthetics standards is changing gradually.

Goldfish keeps evolving unceasingly, arising with so many different beautiful characteristics. Discovery and innovation is the everlasting theme for this culture. For the appreciation on goldfish, there is no best, only better.

风华再现

——台湾地区金鱼市场历程、现状与动向

文 / 黄之旸（台湾）

金鱼，在台湾地区是熟悉却又陌生的观赏鱼种，熟悉的是，那早从宋朝即有的形象，以及分别历经池养、盆养乃至现代水族的缸养，依旧热潮不退的风采；陌生的是，那与时俱进的多变形态、色彩与诸多特征，以及世界各国积极发展的养殖技术与日益严苛的审美标准。台湾在金鱼饲养上，有着超过50年的流行与稳定市场发展，且几乎就是当地的水族市场发展历程，从头至今皆有参与，声势持续不坠；而与其说是观赏鱼，倒不如说是形同犬猫般的宠物，因为广泛的饲养环境、对环境优异的适应能力，以及多变形质特征、与人亲密互动，乃至相对一般观赏鱼明显较长的寿命，皆让金鱼足以称为伴侣动物，并绝对是值得饲主珍宠的伙伴。

踏实地缓步发展

在台湾地区，金鱼的饲养与商业生产发展极早，甚至与锦鲤（koi）一般，皆可视作为台湾地区水族产业发展之滥觞。1940年甚至更早，金鱼便已分别出现在台湾宜兰、彰化及高屏一带，除为一般人家池养，与庭园景观一并呈现欣赏价值之宠物外，同时也开始有小规模的少量生产，而生产种类多以单尾、单色或具简单特色的品系为主。受限于民生经济、社会风气及相对明显的贫富落差，金鱼的饲养难称普及，但对于当时经济正蓄势待发的台湾，能在家中的水缸或庭院里，拥有一池泳动起来娴娜多姿的金鱼，往往是身份地位与经济能力的象征。而当时的金鱼贩卖，则多以小贩挑着担子，沿街或于市集兜售形式为主。

常保活力的市场

华人对金鱼总有股难解的迷恋与爱好，加上风生水起的习惯，以及"视水为财"的观念，因此若是生活条件许可，总希望能拥有一只水缸甚至一方水塘，并让飘逸优雅的金鱼悠游其间。而金鱼又以品系、颜色、形质特征乃至体形，特出于一般观赏鱼之上，因此除了一般选购上的偏好外，在喜庆年节金鱼多是备受欢迎的饲养对象，或是不乏利用文金、蛋金、草金与龙睛等不同品系间形态，乃至颜色的特征差异，搭配成符合个人喜好并具有好运含义的组合。例如丹顶、琉金、珠鳞与五花等，皆是台湾地区极受欢迎的品系。

兼容并蓄的汲取

2000年起，我国台湾地区的金鱼市场，在原本以祖国大陆与泰国为主、本地自行生产为辅的供应态势下，由爱好者引入了日本的系统，这在当时或许仅为满足特定金鱼爱好者的动机，却引发了中、泰与日系金鱼在台湾地区市场的剧烈冲击震荡，原本平静稳定的市场，在短短5～10年间，有了明显的变化。除高质量与崭新品系的中国大陆、泰国与日本金鱼，令本地市场及爱好者大开眼界外，顺势引进的相关信息、名家群雄及其所坚持的饲养方式、选别经验与审美观点，也让台湾地区金鱼市场有了全新与加速发展的趋势，甚至利用网络信息、两岸频繁互动与贸易流通，快速提升了本地对金鱼的认知与品评水平。

等待风起的昂扬

台湾地区观赏水族具有持续稳定的发展，金鱼则在其中扮演了强大的指引功能，而随着新颖品系、专业信息与严苛标准的陆续引进与建立，如今，金鱼饲养在台湾地区，有了形同脱胎换骨般的转变。它或许仍持续在许多庭园、家居或公司行号中，充当映衬景观或呈现喜气的妆点，但在今天还将升华成更为品味与专业的象征，例如每年分别于台湾地区各地，由同好或协会发起的品评赛事，各国同好间的相互参访交流，以及持续引入的新颖品系与实时信息，皆让此刻台湾地区的金鱼市场起首昂扬，风华再现。

The Resurrection of Blossom:
The process, current status, and trend of goldfish market in Taiwan, China

Article / Huang Zhiyang (Taiwan, China)

Goldfish is a familiar and also strange variety in Taiwan, China. It's familiar because the general image has been as it is since Song Dynasty, and also has been raised in pond, in bowls and in modern aquarium; it's strange because of its various morphology, colors and tremendous varying characteristics, as well as abundant breeding methods, stricter and stricter aesthetics standards from all over the world. In Taiwan, China, goldfish has over 50 years' popularity and has developed a stable market, serving as the representative in the entire aquatics development history. Goldfish cultivation never fades in Taiwan but keeps blossoming day by day; People treat it as a cat-like pet rather than a fish for appreciation. With the advantages such as accessible raising, good adaptation to environment, variable characteristic, mildness, and longer lifespan, goldfish is no doubt to be called a mate pet to the raisers.

Steady and Slow Development

The cultivation and commercial production of goldfish began from a very early time. Goldfish and Koi are both the origin for the development of aquatic market in Taiwan, China. From 1940 or even earlier, goldfish appeared in the districts of Yilan, Zhanghua, Gaoping in Taiwan, China. In addition to being kept in ordinary pond, as ornamental landscapes as garden plants, it also had started to reproduce in small scale; but the main varieties were with single tail, single color or other single characteristics. Limited by livelihood economy, social morality, and obvious gaps between rich and poor, the cultivation of goldfish cannot be popularized at that time. Therefore, it was a reflection of welfare and social image to raise a pond of goldfish in the tank or in the garden yard. At that time, goldfish was sold by street vendors or small fare resorts.

The Flourishing Market

Chinese people always have a crush on goldfish. With a faith of "Fengshui", they think water is the symbol of "wealth". Therefore, once it is affordable, people will try to have a tank or a pond to raise goldfish, watching them swim gracefully and freely; Goldfish beats other ornamental fish by its variety, color，feature andshape. They are preferable especially on the celebration occasions. The popular varieties can be exemplified by Fantail Goldfish, Egg-Fish, Common Goldfish and Dragon Eyes. The distinction on shapes and colors when gathered together means good fortunes. Additionally, other varieties like Red Cap, Ryukin, Pearlscale and Calico Goldfish, all popular as well in Taiwan, China.

Inclusive Learning and Introducing

Originally, the Taiwan goldfish market was mainly supplied by Mainland of China or by Thailand, with the supplement of self-producing. Since 2000, some fanciers brought in goldfish from Japan to carter for specific needs, which has greatly impact goldfish serials from Mainland of China, and goldfish from Thailand in Taiwan, China market. Consequently, the market is no longer stable and peaceful and has been dramatically changed in five to ten years. Many high-quality and brand-new varieties have been brought in from China, Thailand and Japan. Together with these new varieties, a great deal of related data, information of famous goldfish keeper and their special cultivation and selection experience are also delivered to Taiwan, extremely accelerate goldfish development of Taiwan, China; People has acquired more knowledge and expertise on goldfish through internet connection, frequent interactions and business trade with China Mainland.

The Blossom Waiting for Wind

The aquatic animals for appreciation in Taiwan, China is developed in steady steps, among which goldfish plays a strong role of introducing. With the introducing and establishment of new varieties, professional information and strict standards, the cultivating of goldfish in Taiwan has been largely improved. In many slogans of garden or housing companies, goldfish may still be used to foil background or represent the meaning of fortune. But nowadays it is more a symbol of taste and profession. For example, every year, in Taiwan areas there are appreciation competitions launched by enthusiasts or communities, communications among enthusiasts from different countries, new varieties and real-time information introduced from every country, all made Taiwan's goldfish market blossom and resurrected.

浅谈金鱼变异

文 / 杨小强

故事开始于"色诱"

18世纪，博物学家卡尔·林奈第一次见到形似产于欧洲的黑鲫，体色却是金黄的金鲫鱼，还以为中国的鲫鱼都是如此艳丽，他将金黄色（*auratus*）做为种加词命名绝大部分个体还是银灰色的鲫鱼（*Carassius auratus*）。

鲫鱼是东亚大陆再普通不过的鱼类，林奈命名的鱼类让几千年来以此为食的中国人表示看不懂，还好会去关心拉丁学名的人只在少数。

再普通的鱼中也会有离经叛道的个体，鲫鱼中偶尔出现的金鲫鱼让林奈先生犯了以偏概全的错误。金鲫鱼的出现是鲫鱼体表黑、红、黄三种色素细胞中的黑色素细胞出现了状况。这种状况会出现在金鲫鱼胚胎期，一种是根本就没有分化出黑色素细胞，另一种是分化出的黑色素细胞徒有其名，没有合成黑色素的能力；也会出现在胚后发育期，原本好好的鱼长着长着黑色素就合成不了了，如有一种称为"包金"的金鱼。不管出现在什么时候，也不管是哪根筋出错，唯一的结果是"黑幕"拉开，红、黄两种色素细胞就暴露了出来。

黑色素细胞的"叛变"在动物界普遍存在，比如人类也有先天的"白化病"和后天的"白癜风"，最常见的是毛囊中的色素细胞将停止产生黑色

素，"嗟我白发，生一何早"，白晃晃的顶在头顶令我辈烦恼。自然界中变异出现的金鲫鱼的烦恼可比人类大多了，失去了背灰腹白的保护色将使被捕食的概率蹭蹭上蹿。

自然界中的金鲫鱼、金鲤鱼、金……不管什么鱼，肯定会出现，但偶有发现，因为出现体色变异的概率极低，也因为生存压力太大，应了那句话：自古红颜多薄命。

基因再次突变，让精彩呈现

金鲫鱼艰难地生活在江湖间，人类的手伸了进来，金鲫鱼的命运迎来了转机。接下来的故事大家都知道，游过了放生池时代、家池时代和盆养时代，1500年间金鲫鱼忘记了曾经的江湖，演变成多姿多彩、需要人类悉心呵护才能生存的金鱼。

这其中发生了什么？根本的驱动力是什么？

陈桢教授在1954年发表的《金鱼家化史与品种形成因素》论文中提出："生活条件的大改变必定在金鱼的发育、生理、形体上产生很多影响。比较明显的是活动空间缩小使鱼的游行缓慢，游行缓慢引起形体与尾部的变化。形体与尾部的变化影响到背鳍与臀鳍的退化。"此类以环境变化引发金鱼形态变化的观点，容易被人接受，却不正确。

现代生物学研究表明，生物的性状受基因控制。作为金鱼生物学和金鱼演化史的研究泰斗，由于受当时的政治氛围的影响，基因学说门下弟子陈桢先生接受了米丘林学说的观点。

金鱼体形变得短圆是控制骨骼发育的基因发生突变，是脊椎骨萎缩融合并向下弯曲在形体上的反映。在大自然"物竞天择，适者生存"的丛林法则下只能被淘汰的废品，在人类的呵护下，在缺乏竞争的盆缸中，获得与正常型的兄弟们同样的生存机会。养鱼者将正常型的金鱼从盆缸中剔除，留下短圆体形的突变者，并通过定向培育将"奇葩"保留了下来。

生物的变异是放散的，而丛林法则只为某种特定的生物留下窄窄的生门，只有扛得住选择压力的生物才能生存。对金鱼而言，在人类的保护下生门扩大了一些，但还要再过一道门，那就是人类挑剔的目光，只有符合人类审美观的个体才不会被挑走剔除。

金鱼为什么容易发生变异？为什么近亲繁殖这么多代金鱼还能存在？许多研究者从不同角度给予解释，我认为最主要的原因是与发生在1500万至2000万年前的一次鲫属鱼类染色体同源多倍化事件——即染色体数由50条翻倍成100条有关。多了一套染色体的鲫属鱼类在繁殖过程中，基因重组时会出现更多样化的基因型，因而有了更大的变异潜力。

多了一套染色体还有一个好处是在近亲繁殖时缺陷基因的暴露概率大大下降，这是盆养时代开始延续至今800多年肆无忌惮的近亲繁殖而金鱼尚未绝种的物质保证。

绕不过的环境影响

基因控制论说得如此斩钉截铁，必遭到金鱼生产者声色俱厉地反驳：还有养功之说吗？

是啊，如果基因能控制一切，培育金鱼岂不是和工厂化生产一般的无趣吗？同一种金鱼在不同养殖人手中养出不同的风格，这是养殖金鱼的乐趣和市场风险所在。

有研究者将一尾琉金金鱼作为母本进行雌核发育试验，培育出染色体组型完全相同的后代，其表现出的模样也不是想象中一个模子倒出来的，它们在体形和体色上出现较大的差异。试验告诉我们，基因也不能完全决定一切！

我用质量性状和数量性状来解释这个问题。

质量性状比较稳定，多由一对或少数几对基因所决定，不易受环境条件的影响，它们在群体内的分布是不连续的。如金鱼的尾鳍，要么是单，要么是双；如臀鳍要么有，要么没有。它们都是在胚胎发育期就定下来的，后天怎么养都休想改变，除非你动了剪刀。如果你非要动手，那得尽早，在胚胎发育的早期通过控制基因的表达来实现，有人这么干过，并取得了成功。

数量性状是指在一个群体内的各个体间表现为连续变异的性状。数量性状为多基因遗传性状，数量性状的变化受基因和环境的影响。在基因影响方面，若干对基因都能控制数量性状，而各个基因产生的效应都很微小，表现出的性状是控制同一数量性状的基因共同作用的结果。在环境方面，养殖金鱼主要影响数量性状的因素有温度、营养、水体等等，尤其在某个观赏性状发育敏感期，营养、温度等环境条件能左右金鱼数量性状的发育，影响金鱼的品质。

金鱼的头茸、绒球、水泡、花色，还有体形等观赏性状均属于数量性状。比如头茸的形态，受多基因控制，头部不同区域表皮中疏松的结缔组织通过分布的厚薄，表现出高头、虎头、狮头等不同类型；受发育敏感期营养的影响，结缔组织发育快则头茸发育充分，你饿着它，头型还是或高头、或虎、或狮，只会象征性地薄薄长一点。

比如体形，试以兰寿金鱼"虎猪之争"论之，背部肌肉发达的好品种（好的基因型）在苗种期给予充足的优质蛋白质（常用的是水丝蚓）、适当的运动，可培育出来虎虎生威的兰寿。如果选择背幅较小的亲本（不那么好的基因型），再给予过量的营养，整天只顾着埋头猛吃不运动，光长肚子不长背，那培育出的就是一只"猪"。

这里还列举了金鱼的花色，金鱼色斑分布受多基因表达影响，环境因素对其影响主要表现在颜色的浓淡，以及变色速度的快慢。

所谓养鱼人"养功"的高低就体现在对不同品种金鱼，在不同观赏性状发育敏感期环境条件的调控水平。

总之，基因决定金鱼的观赏性状的发育，观赏性状的发育程度则受环境因素（生活条件）的影响。

Talk Shallowly about Goldfish Variation

Article / Yang Xiaoqiang

The story begins with "The color allure"

In the 18th century, the first time Naturalist Carl Linnaeus saw the shape of the black carp produced in Europe, but the color was golden. He thought all Chinese *Carassius auratus* would be so gorgeous. He considered *auratus* as an additional word and named most of the individuals are still silver-gray *Carassius auratus*.

Carassius auratus is the East Asian continent fish which can not be more common. For thousands of years, people who take the fish as food could not read the fish's name by Linnaeus. Fortunately, there is only few people care about the Latin scientific name.

Ordinary fish will also have deviant individuals, Because of the occasional occurrence of *Carassius auratus*, Mr. Linnaeus made an over-generalized error. The occurrence of *Carassius auratus* explained that there is something wrong about the melanocytes in three pigment cells of black, red and yellow. This situation will appear during the embryonic period. There are two possibilities, one is that there is no differentiation of melanoma cells at all, the other is that, the differentiated melanocytes only in name don't have the ability to synthesize melanin. This situation will also occur during the post-embryonic period, at the beginning the fish grows normally, and then it can not synthesize melanin, just like the fish named "wrapped gold". As for the "wrapped gold", whenever the situation occurs and no matter what is wrong with it, it will end up with the appearance of red and yellow pigment cells and the split of the black.

Melanocytes "mutiny" is widespread in the animal kingdom. The human race for example also has congenital albinisms and postnatal albinisms. What the most common is that the pigment cells in hair follicles will stop producing melanin, "people sigh, how early that we have got the white hair" and the hair is so white on our head that bring us so much annoyance. The trouble of golden carp appear by variation in nature is so much bigger than that of human beings. It is more difficult to prey on without the protective colors: gray on the back and white on the belly. In the opposite it is much easier to be prey.

In nature no matter which fish, for example, golden carp, golden cyprinoid, and so on, will definitely occur. Because the probability of color variation is very low and there is too much pressure to survive, the consequence is just like the saying: Pretty born unlucky.

Gene variation again, comes the wonderful presentation

Golden carp have lived a very hard life until human beings take their actions when the fate of golden carp ushered in a turning point. The next story we all know that golden carp have gone through several stages, pond where fish are released stage, home pool stage(the golden carp were kept in a pool at home) and pots stage. (the golden carp were kept in pots). During 1500 years, the golden carp had forgotten its fish world in the past. It have evolved into a goldfish that is a variety of colorful and needs the care and devotion of human beings to subsist.

What happened? What is the fundamental during force?

In the thesis *Goldfish Domestication History and Breed Formation Factors*, issued in 1954, professor Chen Zhen mentioned that, "the great changes in living conditions must greatly influent the development, physiology and physique of the goldfish. It is obvious that the reduced space for its activities makes its swimming slow, therefore, the physique and tail part will change, which will influent the degeneration of the dorsal fin and anal fin. "The point of view that, the changes in the environment caused the changes in goldfish morphology, is easy to be accepted, but not correct.

Modern biology research states, the properties of living organisms are controlled by genes. As a leading authority on goldfish biology and gold fish history research, as a result of being influenced by the political atmosphere at that time, Mr. Chen Zhen, a gene theory disciple, accepted the theory of Ivan Vladimirovich Michurin.

The goldfish turns into short and round which reflects that, the genes controlling skeletal development have mutated and the spinal bone has been atrophied and deflexed. In nature, it can only be eliminated under the jungle law, that the survival of the fittest, while it will obtain the same living chances with other normal fish brothers in a jar without any competition protected by human beings. Fish breeders remove normal goldfish from bath and keep those short and round body of variation, which will be oriented cultivated to be "wonderful" retained.

Biological variation is diffuse, and the law of the jungle leaving only a narrow green door of a particular organism, only those organisms which can resist selection pressure will survive. As for the goldfish, although the green door for living chances is wider with the protection of human beings, but there is another door, people's critical eyes. The only individuals safe from elimination are those are aesthetically pleasing to human beings.

Why do goldfish prone to variation? Why they can still mutate since so many generations of inbreeding? Many researchers from different angles to give explanations. In my opinion, the main reason has something to do with an event occurred in 1500 to 20 million years ago, a *Carassius auratus* fishes homologous chromosomes are polyploid, that is, the chromosome number is doubled from 50 to 100. Added one set of chromosomes, during its breeding process, the gene recombination will result in more samples of the genotypes hence it has a greater potential for variation.

With one more set of chromosomes there is another advantage, that is, during inbreeding, the exposure chances of defective gene are greatly reduce. Pot raising ear until today, 800 years of unbridled inbreeding, it guarantee in material that the goldfish has not yet extinct.

Impact of the environment

Gene cybernetics put it so categorically, and the goldfish producers must be stern retorting: is it still has anything to do with the breeding practices?

Yes, If genes can control everything, isn't it true that goldfish cultivation would be so boring as factory production? The same type of goldfish will give birth to different styles in the hands of different breeders, which is the joy of goldfish breeding and also the risky part of the business.

Researchers have a Ryukin goldfish as a female parent in gynogenesis test, cultivating identical karyotype offspring but the appearances are not the same as in the imagination. They appear different in size and color. The experiment tells us that the gene can not completely determine everything.

I explain the problem by quality and quantitative traits.

Quality traits are relatively stable, which are determined by one pair or a few pairs of genes, and are not easy to be affected by environmental conditions. Their distribution in the population are not continuous, Such as goldfish caudal fin, either single or double; goldfish anal fin, either have or not. This is all decided during the growth phase of the embryo and it is no way to be changed postnatal except using scissors. If have to, take actions as early as possible. During the early growth phase of the embryo, change it by controlling the genes. Some people have done so, and achieved success.

Number of characters means that each individual in a population continuously mutates. Quantitative as a multi gene genetic trait, the changes of quantitative characters were affected by the gene and environment. In the aspect of gene effect, several genes can control the quantitative traits, and the effects of each gene are very small. The trait appearance is the result of the common action of genes controlling the same quantitative. In the aspect of environment, the main factors affecting the number of goldfish breeding traits are temperature, nutrition, water quality and so on, especially the nutrition and temperature during a sense of development period of ornamental characters will influent quantitative traits development and the quality of goldfish.

Goldfish ornamental traits such as head pompon, pompons, Bubble-Eye, color, size and so on, are all quantitative traits. For example, Due to multiple genes' control, according to the distribution thickness, loose connective tissue in different regions of the head shows different types: Tall-Head, Tigerhead, Lionhead, and other different types. Affected by the nutrition during sensitive development period, if the connective tissue grows fast, the head pompon will fully-develop; If there is not enough nutrition, its head will also be Tall-Head, Tigerhead, or Lionhead, or symbolically grow a little which is very thin.

Such as size, try to start with Ranchu goldfish "Tiger pig dispute", a good variety (genotype) with muscular on the back if there is adequate protein in excellent quality (commonly is water earthworm), and proper exercises can be cultivated to imposing and vigorous Ranchu. If we give excess nutrient to a parent with small back (not so good genotype) and it is busy in eating without any exercise, belly growing but back, then we will cultivate a "pig".

Here is also a list of goldfish colors, color stain distribution expression affected by multiple genes; Impacts from environmental factors are mainly manifested in color shading and color changing speed.

The so-called fish breeder "breeding skills" is reflected in the ability of regulating environment condition for different goldfish varieties, during different ornamental traits growth sensitive period.

In short, gene determines goldfish development of ornamental traits, which is influenced by environmental factors (living conditions).

参考文献
References

王春元. 金鱼的变异与遗传 [M]. 北京：中国农业出版社，2007.

袁珂. 山海经校注 [M]. 上海：上海古籍出版社，1983.

沈伯平. 金鱼文化艺术欣赏 [M]. 扬州：广陵书社，2014.

叶金明. 赏心悦目：扬州金鱼文化 [M]. 扬州：广陵书社，2014.

岳珂. 桯史. 北京：中华书局，1981.

陈桢. 金鱼家化史与品种形成的因素 [M]. 北京：科学出版社，1955.

李明德. 鱼类分类学 [M]. 天津：南开大学出版社，2013.

陈帧. 金鱼的家化与变异 [M]. 北京：科学出版社，1959.

王春元. 金鱼的变异与遗传 [M]. 北京：中国农业出版社，2007.

文形

七画

八画

九画

蛋形

七画

八画

九画

十一画

十二画

文形

蛋形

Fantail Goldfish

Egg–Fish

C

S

T

V

W

后记

文 / 林海

作为一名金鱼爱好者，一直希望拥有一部品质上佳的金鱼图鉴。当我初为人父，为孩子抱回一部部"DK百科"时，这个愿望变得更为强烈。没想到，梦想这么快就照进了现实，更没想到的是，我也有幸参与到"烘焙"梦想的过程中。

眼前这部《中国金鱼图鉴》，在我看来，至少有三方面价值。

其一，欣赏价值。金鱼是"活的艺术品"，艺术品于人，最本然的好处便是赏心悦目。金鱼自古有"凌波仙子""水中奇葩"等等的美誉，表达了人们对美的追求与陶醉。但金鱼也有它的"麻烦"——活的。活的，便意味着易朽，这让金鱼不能如文玩清供赢得近乎"永恒"的瞻观——依稀记得小时候在自然博物馆里确乎是见过金鱼标本的，但福尔马林泡出的物件，实在让我无法与金鱼之美作过多的联想。于是，我们借助影像的定格。一张张瞬间的捕捉，让一尾尾活的生命仿佛超越了自身，进入了"永恒"。当然，这里并不想探讨摄影的艺术特性，而意在表达图像对于金鱼格外的意义——如果没有图像，一条金鱼之美的辐射半径，恐怕只能是"此时此地"。进而想说的是，虽然国内的金鱼图鉴出过不止一部，但眼前这部依然堪称个中翘楚，这得益于展示鱼只遴选之精当，也得益于图像品质甄别之严格。艺术之金鱼与摄影之艺术粲然交汇，虽不能谓之完美，但其间的用心和品位历历可感。所以，亲爱的读者，当你面对这部《中国金鱼图鉴》，实则也面对着一部艺术欣赏图典。或许有朋友会说：我就是看不出金鱼哪儿美，怎么办？那好，请抱着这本书再翻读三遍！

其二，史料价值。迄今为止，对中国金鱼演化史的探究，多跳不出半个多世纪前陈桢教授所著《金鱼家化史与品种形成的因素》一书。在叹服陈桢先生钩沉索隐、苦心孤诣的同时，我们也不能不为他感到遗憾，毕竟，中国金鱼史料是如此之少，其中对金鱼品种的描述又多语焉不详。好不容易出了部《金鱼图谱》，却印得墨彩糊涂，难以明辨。文字与图像，决定了我们能否对一个金鱼品种出现年代做出准确的判断，进而勾连出金鱼系统演化的图谱。可惜，这幅图谱恐怕要永久性地悬疑了。近代以来，摄影术的普及让金鱼"修史"有"像"可寻。更有老前辈如傅毅远、伍惠生先生，不但编制了《中国金鱼品种名录》，而且标注出每个品种的出现年份。这无疑是开创性的工作，但也如傅先生所说，此名录"调查范围仅限于杭州一处，遗漏和错误之处，尚请各地金鱼专业饲养单位和业余爱好者予以指正充实，以期完善为感。"诚哉斯言！非傅先生不愿勉力而为，实是信息传递条件所限。所以，窃以为金鱼"修史"的新时代，当以数码摄影和互联网的发达为标志，由此看这部《中国金鱼图鉴》的出版，可谓应运而生，正当其时。另一方面，据过往的阅读经验，我知道每部图鉴，对一些金鱼品种都有些"立此存照"的意味。因为金鱼演化的速度是如此之快，可能几年之间，一个品种就会黯然退场。翻看金鱼老图片，我常会感叹："原来那时还有这个品种"——但愿若干年后再看这部图鉴时，我们能多些欣悦，少些怀念。

其三，认知价值。中国金鱼科普的一大难题，是分类和命名的混乱。就分类而言，三分法（文、龙、蛋或草、文、蛋）、四分法（文、草、龙、蛋）还是五分法（文、草、龙、蛋、龙背）莫衷一是，但各类之间区分标准不统一，逻辑上禁不起推敲。此部图鉴，只以背鳍有无作为唯一的尺度，两分为"文形"和"蛋形"，一以贯之，一目了然。在分类之下，又以性状变异（包括体色变异和形态变异）作为品种判定的标准，进而，推演出"两段式"的品种命名法，即"体色变异+形态变异"如"五花水泡"。过往"多段式"的命名法，基本可概括为"体色变异+特征1+特征2……形态变异"，如"五花长尾软鳞水泡"。但如果"特征"未达到品种的高度——或遗传不够稳定，如"玉顶""十二红"等；或已成为各品种的普遍性特征，如"四尾""长尾"等——随意添入品种的命名中，极易造成混乱和误解。"二分法"和"两段式"，删繁就简，还原出一幅逻辑清晰的中国金鱼品种分支图。这看似基础的工作，实则对金鱼审美、金鱼培育以至金鱼产业的发展等都将产生深远的乃至是革命性的影响。至少，有助于我们从眼花缭乱的"前缀"中找到进出的路径，同时，让我们对"品种繁多"少些不假思索的自负和迷恋。值得一提的是，本书首次将金鱼头部变异分为"虎""狮""龙""鹅"四种，这对解决长期困扰国人的"虎头""狮头"因地域文化多元而导致的同鱼异名或同名异鱼的问题，提供了一把钥匙。只是为了尊重习俗，本书保留了当下通行的称谓："猫狮""寿星""兰寿"等。相信若干年后，"虎""狮""龙""鹅"将成为中国金鱼新的"习俗"。

以上所述，不乏开创意义的思考和探索，特别是金鱼品种的分类和命名方式，相信会引起不少业内方家的讨论，激赏或非议，都是正常的——不说别人，本书的编委们也曾争论得不亦乐乎。最终，我们把它呈现在读者面前，一则抱着正本清源的信念，二则以开放的态度期待读者的批评和指正。即使是在"金鱼的品评标准"这类更为主观性的领域，我们也不会因为"见仁见智"而怯于表达自己的观点，正如在"金鱼的演化历史""金鱼的南北分布"等相对客观性的领域，我们也绝不认为自己的陈述无懈可击。因为我们的心愿，就是集众人之心力与所长，奉献一部与中国金鱼的发展面貌和文化内韵相匹配的图鉴，我们会尊重写下的每个字、选定的每张图。

也因此，特别感谢海峡书局提供这样一个良好的平台，让来自金鱼界的编委们能够完成自己的，也是千千万万金鱼爱好者的心愿。据海峡书局林彬社长介绍，这部《中国金鱼图鉴》是海峡书局继《中国鸟类图鉴（全3册）》——曾获第五届中华优秀出版物奖图书奖——出版的又一部自然生态图鉴。海峡书局有志于策划、出版中国的生态百科系列，成为中国的DK。这的确是富于远见卓识的创举！而身为两部图鉴主持者的曲利明副社长，既是出版家也是摄影家，在即将"解甲归田"的年纪，还能以不倦的热忱，抱着"一旦决定做，就要做到极致"信念来打造中国的博物类图书品牌，诚然令人感佩。

"我们从拍摄身边的物种开始，让大家对它们有个感性认识，热爱它们，保护它们，建立起人和自然和谐关系。"曲利明先生的话虽朴素，却道出了现代博物学的真谛。在我看来，当下博物学的复兴，正是一个百年间经历了几度文化劫难的民族启动的"自我修复"。这是"爱的教育"，是对"天地人神"的反思和领悟，是对富有人性的生活世界的回归。而金鱼，作为人文化育的物种，对它的观照，是人与自然的对话，更是人与文化的对话。因此，我认为这部《中国金鱼图鉴》，所做的是对自然的致敬，更是对文化的传承，是"博物"之上的"博物"。进而，我还期望这部图鉴，能够每三至五年更新一版，不仅因为金鱼的演化速度之快有别于自然物种，更因为我们能从演化背后，读出自然与造物的灵魂，读出文化与人心的脉动。如果说编辑是"遗憾的艺术"，那么，就让我们在一次次的更新中去接近完美——永无止境。

写到这里，我不禁更加期待这部《中国金鱼图鉴》的出版——海峡书局"中国自然生态系列丛书"第二部，一只新出炉的色质香浓的面包——我会把它抱回家，让它成为儿子的床头书。

Postscript

Article / Lin Hai

As one of the enthusiast for goldfish, I have been hoping to own a nice book on goldfish for a long time. After my son was born, as a father, the wish to give my kid a serial book like DK encyclopedia has become much stronger. It is unbelievable that the dream has been realized so quickly, and I myself, could even be part of the dream maker.

The Goldfish of China is at least of three values for me.

First of all, appreciation value. Goldfish is a living art. The essence of an art is for appreciation. Ancient depictions like "faery in water", "gorgeous water flower" and so on are conveying people's affection and chasing for beauty. But Goldfish has a problem due to "live". Live means "to be dead"; Goldfish cannot serve as a permanent ornamental object as others articles for amusement. I remember when I was still young, I had a chance to see goldfish specimen on the exhibits in the museum, and the dead goldfish after being processed by formalin definitely could not let me connect with "beauty". Therefore, people have chosen to take photos to record wonderful moments of moving goldfish. In that way, goldfish seem to be alive again and become immortal to some extends. Of course, photographing is not the key topic here, but I would like to express the special significance of it to goldfish-- without photos, the beauty of a goldfish can only be displayed in limited timing and space. Although there are many similar books of goldfish with illustrations, the one at hand is the best of them, not only for its prudent selection of the varieties, but also for exquisite quality of the photos. The combination of art of goldfish and art of photographing has perfectly embodied the unique taste and style of the editors. So dear readers, reading this book is like reading an art of atlas. Some friend might say, I don't see the beauty of the goldfish, well, please read this book three more times.

Second of all, historical reference value. So far until now, many researches on Chinese goldfish have been framed with *Goldfish Domestication and Elements for Variation* by Professor Chenzhen 50 years ago. Although we are impressed by Mr. Chen's great annotation and painstaking study on goldfish, we still feel something is missing, for example, there was limited historical records and obscure descriptions on goldfish varieties. *Goldfish Illustration* was treasured for a while, however, it does not meet people's satisfaction at the end because of its vague printing. Through words and images, people can tell precisely the time that a specific a variety appears, and then make a serial connection of the entire goldfish atlas. Pity for *Goldfish Illustration*, the chance of realizing this goal would be very slim. Since modern times, photographing has made it possible for goldfish to be presented not only with "words" but also with "images". Senior predecessors like Mr.Fu Yiyuan, Mr.Wuhui, have edited the *Index for Chinese Goldfish* with detailed annotation for goldfish creation year with pictures. The book has been known as a milestone at that time. However, it has restrictions too. As is put by Mr. Fu, the research in this book is mainly limited to goldfish in Hangzhou area, it might have inaccurate descriptions, and it is open to receive corrections or supplements. It is really true! Actually, Mr. Fu had tried hardest. However, it was too restricted on information communication at that time. So it is just the right moment for this book to be published for goldfish history revision under a climate of developed photographing and internet. In addition, the evolution of goldfish is so fast that after several years, one variety might disappear. With this book, by looking at the those atlas I can recall that this variety has existed before: I wish that many year later, when we read this book, the mood is pleasant other than regret.

Third, cognition value. One of the most difficult topics on goldfish is standardization of classification and naming. As for classification, there exist trichotomy (Fantail Goldfish, Dragon, Egg-Fish or Common Goldfish), and quartering (Fantail Goldfish, Common Goldfish, Dragon and Egg-Fish), or quinquepartite method (Fantail Goldfish, Common Goldfish, Dragon, Egg-Fish, Dragon with High Hump). But none of them has established a clear classification standard However, this book has uniquely set up a new way of classification, that is with or without dorsal fin, goldfish can be divided into two types" Fantail Goldfish " and "Egg-Fish", clear and consistent. As a subsidiary, goldfish can also be classified per color variation or morphology variation, which generates the dichotomy naming system, like "Calico Bubble-Eye", it is named per "body color variation+ morphology variation". In the past, there was also multi-part naming system, which was based on color variation+feature1+feature2... such morphology variations and so on, but when the feature is not stable enough or not capable to be called a new variety, e.g. "Jade Head" or "Twelve Red"; or when the feature has gradually become common for all other varieties, e.g. "Four Tail" or "Long Tail", this type of naming method would cause confusions and misunderstandings. Dichotomy and two-part naming has helped to set up a clearer family tree for goldfish of China. Although it seems only a basic task, it indeed has brought with a profound and evolutionary influence on goldfish appreciation, breed cultivation and even the industrial development. With the new naming, it is very easy to recognize a variety and stops us from blind pride for "tremendous varieties": they are actually same variety with different naming. It is worth noting that this book has initially classified the head variation with Tigerhead, Lionhead, Dragonhead and Goosehead, which has solved the problem of mixing up by the names of varieties due to regional culture difference. Some popular naming like Lionhead (with Well Developed Head Growth and Chubby Body), Lionhead (Southern), Ranchu are still kept in this book, but only as a respect for tradition. We believe that after a few years, suffix like" Tiger", "Lion", "Dragon", "Goose" will become a new tradition for naming Chinese goldfish.

In conclusion, this book is of great innovation and exploration significance, especially on goldfish classification and naming. It is believed to create lot of disputes among experts professional. However, it is quite normal either to be appreciated or to be criticized. Even among the editors ourselves, we have been debating all the time. There are two principles to guide us: to set up right faith on one hand, and to receive corrections and criticism on the other hand. For example, we won't be timid on expressing our own idea on, for example goldfish classification although it is a very subjective item. We would never think our descriptions are perfectly right when it comes to "goldfish evolution history" and "goldfish allocation in North and South". To be honest, with collecting strong points of the masses, our target is to provide you an atlas that has absorbed the core value of local cultures. We respect every single word and every single picture we have put in this book.

Hereby, special thanks to The Straits Publishing & Distributing Group (SPDG for